Cave and Karst Systems of the World

Series Editor

James W. LaMoreaux, P. E. LaMoreaux and Associates, Tuscaloosa, AL, USA

This book series furthers the understanding of cave and karst related processes and facilitates the translation of current discipline-specific research to an interdisciplinary readership by dealing with specific cave or karst systems. Books in this series focus on a specific cave or karst system, on the cave or karst systems of a specific region, on a specific type of cave or karst system, or on any other perspective related to cave and karst systems of the world. The book series addresses a multidisciplinary audience involved in anthropology, archaeology, biology, chemistry, geography, geology, geomorphology, hydrogeology, paleontology, sedimentology, and all other disciplines related to speleology and karst terrains.

Ali Yamaç • Ezgi Tok

Caves and Karst of Küre Mountains National Park, Turkey

Springer

Ali Yamaç
OBRUK Cave Research Group
Istanbul, Türkiye

Ezgi Tok
Istanbul Technical University - Eurasia Institute of Earth Sciences
Istanbul, Türkiye

ISSN 2364-4591 ISSN 2364-4605 (electronic)
Cave and Karst Systems of the World
ISBN 978-3-031-80141-9 ISBN 978-3-031-80142-6 (eBook)
https://doi.org/10.1007/978-3-031-80142-6

This Springer imprint is published by the registered company Springer Nature Switzerland AG
The registered company address is: Gewerbestrasse 11, 6330 Cham, Switzerland

Contents

About the Authors

Ali Yamaç is a Turkish speleologist. He was born in 1957 and has been caving since 1976. Over those years, he had surveyed and explored hundreds of caves, some of which are Turkey's longest and deepest caves. He was the President of the Speleological Federation of Turkey during 2007–2009 and the President during the establishment of three different caving organizations. He was the leader of the team that had prepared the Cave Inventory of Turkey. In addition to his natural cave explorations, as a part of his artificial cave projects, he explored and surveyed numerous rock-cut dwellings and underground cities around different regions of Turkey. Starting with the underground structures of Hagia Sophia Church and Topkapı Palace in Istanbul, his artificial cave projects continue with underground structures inventory projects in Kayseri and Gaziantep provinces. Both projects had been accepted to the UNESCO World Heritage Site Tentative List. He had explored and surveyed several rock-cut settlements and churches within the "Euphrates River Cliff Settlements Project" and "Kura River—Ardahan Region Survey Project."

Ezgi Tok is a Turkish speleologist. Her research is focused on microbial life and the related components of cave environments. Since 2016, she explored several karstic caves all over Turkey to reveal the microbial diversity, the ecosystem functioning of microbial populations, and their relation with the psychochemical conditions within the cave ecosystem. Her research also investigates the impact of surface atmospheric conditions on subsurface microclimate. The project she currently conducts is the first cave microbiology study in Turkey that comparatively examines caves from different climatic regions with different environmental conditions. She had been caving since 2009 and joined several national and international speleological expeditions in Turkey and abroad, such as the expedition Lion of Crete, the second deepest cave in Greece, and the expedition of Keş and Çem sinkholes, the sixth and tenth deepest caves of Turkey, respectively. She had surveyed and mapped several caves of Turkey; among them İnsuyu Cave, which is the longest, and Kocain Cave, which is the largest chamber of Turkey, are especially important. She had also joined in artificial cave projects, the underground structures inventory projects in Kayseri and Gaziantep provinces.

Abbreviations

ADEKS	Association d'Exploration des Karsts Subtropicaux
AKUMAK	Akdeniz University Caving Club
ASPEG	Anatolian Speleology Group
BCRA	British Caving Research Association
BÜMAK	Boğaziçi University Speleological Society
BUMAD	Boğaziçi International Speleology Association
CAP	Cide Archaeological Project
FACC	East Anglia Fell and Cave Club
FAO	Food and Agriculture Organization of the United Nations
FSE	European Speleological Federation
HGM	General Directorate of Mapping
GEF	Global Environment Facility
ITUMAK	Istanbul Technical University Speleology Club
LUSS	Leicester University Speleogical Society
MAD	Cave Research Association
MTA	General Directorate of Mineral Research and Exploration
OBRUK	Obruk Cave Research Group
TMF	Speleological Federation of Turkey
UIS	International Union of Speleology
UNDP	United Nations Development Program
WWF	World Wildlife Fund

1 Introduction

This book is a first in its own way. As a result of more than 2 years of collaboration with the General Directorate of National Parks, a comprehensive cave inventory of a national park has been meticulously compiled for the first time in the nation's history. This offers a captivating glimpse into the hidden depths of one of Turkey's ecological marvels.

Nestled between the provinces of Kastamonu and Bartın in Northwestern Turkey, Küre Mountains National Park stands as a testament to the awe-inspiring beauty of nature's craftsmanship. The expansive landscape unfolds like a vibrant tapestry, adorned with towering canyons, cascading waterfalls, and a rich tapestry of unique vegetation. The park's pristine wilderness is home to a diverse array of wildlife, which further enhances its allure and contributes to its reputation as a sanctuary of biodiversity.

It is noteworthy that within the context of Turkey's diverse natural landscapes, which currently host 48 national parks and over 200 nature reserves, Küre Mountains National Park shines as a beacon of excellence. Its steadfast commitment to international environmental standards sets it apart, underscoring its unwavering dedication to conservation and sustainable management practices.

The park's distinctive character is derived from its strategic geographical location, situated on a high plateau encompassing an area of 370 km^2 and extending to an additional buffer zone of approximately 800 km^2. The elevated plateau, which is inaccessible except for one or two points, assumes the character of a pristine wilderness, untouched by human influence. This natural fortification serves to protect the park from the adverse environmental impacts typically associated with human settlements, rendering it an ecological sanctuary. Consequently, the top of the plateau is devoid of any human influence and is home to an astonishing array of wildlife. Here, amidst the network of caves and caverns, lies an ecosystem teeming with untold biodiversity, awaiting discovery and documentation.

The genesis of exploratory endeavors within Küre Mountains National Park traces its roots back to the 1980s, with formal survey studies gaining momentum in 1995. The fruition of these efforts culminated in the inception of the "National Parks and Protected Areas" project in 1997, paving the way for the official designation of Küre Mountains as a national park in July 2000. In the pages that follow, the intricacies of this environmental sanctuary will be unraveled, shedding light on its ecological significance and the imperative need for its preservation in the global arena of biodiversity conservation.

In the rich tapestry of global biodiversity and geological wonders, a multitude of national parks stand as testaments to the extraordinary karst geomorphology that characterizes our planet. Internationally celebrated examples, such as Carlsbad Caverns National Park in the USA, which boasts 119 caves (Carlsbad, n.d.), and Malaysia's Mulu National Park, which is home to colossal cave galleries and the world's largest cave chamber (Mulu, n.d.), underscore the global significance of subterranean landscapes. This phenomenon is observed across diverse nations, including Croatia, Serbia, Switzerland, and France, where designated "cave national parks" have been established. These parks exemplify a shared commitment to the protection and promotion of these intricate geological formations.

In alignment with this global ethos, Küre Mountains National Park emerges as a distinctive entity. Rather than seeking rivalry with renowned counterparts, it carves its niche with the exploration and documentation of 125 caves. The profound significance of this wealth of subterranean wonders prompted the inception of a scholarly endeavor—the creation of a pioneering "Cave Inventory Book," a testament to the park's dedication to unraveling the mysteries concealed beneath its rugged topography. This monumental task, carried out under the auspices of the National Parks Directorate, materialized through an exhaustive cave research project spanning the years 2008–2010, forming the cornerstone of the comprehensive inventory that follows.

The cave inventory of Küre Mountains National Park is a collaborative achievement, drawing on extensive data

A. Yamaç, E. Tok, *Caves and Karst of Küre Mountains National Park, Turkey*, Cave and Karst Systems of the World,
https://doi.org/10.1007/978-3-031-80142-6_1

accumulated over years from diverse caving clubs. Of these contributors, the Boğaziçi University Caving Club (BUMAK) is of particular note, along with other speleological teams. Their indispensable role in the exploration and surveying of 65 caves within and around Küre Mountains National Park reflects the collaborative spirit that underpins scientific exploration (Albukrek, 1986, 1993a, 1993b; Ülkümen, 1983).

During the course of our project, an additional 49 caves were discovered and explored, bringing the total number of known caves to 125 with the contributions made since then.

Yet, amidst the celebration of this monumental achievement lies the acknowledgment of the vast unknown, beckoning further research and exploration. Despite the discovery and exploration of 125 caves, the sheer enormity of the limestone expanse suggests that the actual number of caves may far exceed the documented count. This realization underscores the imperative for continued scientific inquiry and exploration, ensuring a more comprehensive understanding of the subterranean realm of the Küre Mountains.

In the pursuit of unraveling the mysteries concealed within the Küre Mountains National Park, the inherent challenges posed by the plateau's vegetation underscore the necessity of expanding the scope of research beyond the confines of the national park and its buffer zone. Consequently, the inclusion of several significant caves from neighboring districts, such as Amasra and Cide, became imperative for a more comprehensive understanding of the region's subterranean landscape.

The geological composition of Turkey, comprising rocks such as limestone, dolomite, and gypsum, lends itself to extensive karstification, constituting over 40% of the country's total area of 783,562 km^2 (Nazik, 2004; Nazik et al., 2019). Although gypsiferous regions do occur in Central and Eastern Anatolia, the majority of the country's karstic regions are composed of limestone terrains. These limestone regions differ in age, but they are nevertheless widespread. The North Anatolian Mountains and the Taurus Mountains in the south are home to a significant concentration of karst terrain. Nevertheless, a variety of studies have drawn attention to the Taurus Mountains because this mountainous region has several thick, pure limestone layers with a high lime content. Due to factors such as elevation, tectonic features, and the prolonged persistence of karstification, the karstic forms in this area are incredibly diverse and rich.

While historical explorations have predominantly focused on the Western Black Sea, Western Anatolia, and the Taurus Mountains, recent research has unveiled the previously unrealized karstic potential of the Black Sea Region, challenging conventional assumptions about the distribution of karst formations (Aylar et al., 2018).

It is estimated that there are tens of thousands of caves in the karst regions of Anatolia. However, the nascent state of speleological studies in Turkey, commencing only in the 1960s, has resulted in the exploration and surveying of only a fraction of these caves. According to the accessible information, 3,050 caves were included in the TAY Project Cave Inventory of Turkey (www.tayproject.org) (Gürcan et al., 2006). This figure is considerably less than the number of caves in many countries that are smaller than Turkey. For example, it is well documented that there are over 10,000 caves that have been inventoried in countries such as Italy, Romania, Switzerland, and Slovenia, with 8,500 in Greece (Yamaç et al., 2021).

Research on caves in Turkey has predominantly focused on the western Black Sea, western Anatolia, and the Taurus Mountains, where the majority of explored caves are located. In contrast, the Eastern Black Sea and Central Anatolian regions, which are less conducive to cave formation, exhibit a limited number of caves. This emphasizes the need for broader scientific inquiry. The Southeastern Region, which is characterized by extensive limestone outcrops, emerges as a promising area for future cave exploration. This necessitates focused scientific attention.

The critical importance of Küre Mountains National Park as a naturally preserved area amplifies the urgency for new scientific research on its caves and underscores the imperative to integrate the region into a comprehensive conservation protocol, thereby elevating its global significance. However, this commendable pursuit is not without its challenges, as numerous caves across Turkey face serious threats, including unregulated tourism—a perilous activity that poses risks comparable to illegal treasure excavations and habitat conversion.

In contemplating the potential consequences of unregulated tourism, it is crucial to recognize the irreversible damage that can be inflicted upon caves, intricate formations sculpted over millions of years. Chapter 4 elucidates the profound ecological value of these caves, serving as habitats for a diverse array of new and unique species, some of which are still being discovered. The destruction of these delicate ecosystems for tourism purposes not only imperils the rich biodiversity they harbor but also undermines the scientific knowledge that can be gleaned from these pristine environments. Therefore, a judicious and environmentally conscious approach is imperative, emphasizing the delicate balance between promoting tourism and safeguarding the integrity of these invaluable natural wonders.

The subsequent chapters of this inventory book delve deeper into the exploration of the 125 known caves in Küre Mountains National Park, featuring detailed descriptions, maps, and photographs.

After this brief introduction, Chap. 2 is structured with four distinct sections. The initial section provides an overview of Küre Mountains National Park, delineating the region's geographical, natural, and botanical features in intricate detail. The subsequent section, devoted to the geological exposition of the area, departs from conventional scientific discourse, opting for an accessible style free from technical jargon. This deliberate approach aims to make the geological information more comprehensible to a broader readership, employing an informal narrative style. The third section of Chap. 2 delves into the archaeology of the region, with a specific focus on cave archaeology.

In the final section of this chapter, we describe the history of cave exploration in the area, citing primarily Prof. Cemal Arif Alagöz (Alagöz, 1944). Alagöz, one of Turkey's most distinguished geographers, conducted the first comprehensive research in the region. His style is informal yet insightful, and it is reflected in the narrative, providing readers with a rich historical perspective.

The core of this book is found in Chap. 3, which covers 45 of the 125 explored caves within Küre Mountains National Park. Despite the limitations of incomplete reports and maps, the chapter presents documented and photographed caves, prioritizing significance over quantity. The importance of certain caves, even if accompanied by photographs of varying quality, has been acknowledged, and they have been included in the book.

Following a general definition, Chap. 4 discusses the biospeleology of the area and provides insights into the unique ecosystem discovered in these caves. The chapter highlights the biodiversity of the area, showcasing photographs of notable endemic species and providing a comprehensive list of 200 specimens found across 25 different caves during the course of the project. This offers valuable insights into the subterranean life within Küre Mountains National Park.

The conclusion, found in Chap. 5, explains the structural challenges facing the national park and presents recommendations for potential solutions.

The book adheres to the BCRA cave map standard throughout, ensuring consistency and accuracy in the representation of cave systems. Readers seeking deeper insights into cartographic methodologies may consult the detailed explanation of this standard in the back matter. Additionally, a glossary is provided in the final section of the book, elucidating technical terms used throughout the narrative.

This publication represents a significant milestone in the exploration of Küre Mountains National Park. It is acknowledged that the park is an ongoing area of interest, with the anticipation of numerous caves yet to be discovered. We express a collective commitment to fostering further research and exploration in the ensuing years. Moreover, the book aspires to serve as an instrument of awareness, cultivating a profound appreciation for the caves of Küre Mountains National Park and advocating for their preservation, recognizing the inherent importance of safeguarding these natural wonders for future generations.

References

Alagöz, C. (1944). *Türkiye Karst Olayları Hakkında Bir Araştırma (Une étude sur les phénomènes karstiques en Turquie)*. Ankara.

Albukrek, M. (1986). Kastamonu Bölgesi Araştırmaları II. *DELTA, 2*, 5–15.

Albukrek, M. (1993a). Kastamonu Bölgesi Araştırmaları III. *DELTA, 3*, 29–43.

Albukrek, M. (1993b). Kastamonu Bölgesi Araştırmaları IV. *DELTA, 4*, 15–25.

Aylar, F., Zeybek, H., & Dinçer, H. (2018). Karstification and Karstic shapes between the rivers Devrekâni and Aydos of the Küre Mountains. *Eastern Geographical Review, 23*(39), 1–24.

Carlsbad. (n.d.). Carlsbad Caverns National Park. Retrieved July 28, 2023 from https://www.nps.gov/cave/index.htm

Gürcan, G., Yamaç, A., Kırlangıç, S., Pelen, M., Talay, Z., & Zorlu, P. (2006). *Türkiye Mağaralar Envanteri*. Ege Yayınları.

Mulu. (n.d.). Mulu National Park World Heritage Area. Retrieved July 14, 2023, from https://mulupark.com/

Nazik, L. (2004). The karst regions of Turkey (according to morphogenesis and morphometric properties). Paper presented at the International Symposium on Earth System Sciences, İstanbul.

Nazik, L., Poyraz, M., & Karabıyıkoğlu, M. (2019). Karstic landscapes and landforms in Turkey. In C. Kuzucuoğlu, A. Çiner, & N. Kazancı (Eds.), *Landscapes and landforms of Turkey* (pp. 181–196). Springer.

Ülkümen, O. (1983). Kastamonu Bölgesi Araştırmaları I. *DELTA, 1*, 1–9.

Yamaç, A., Gilli, E., Tok, E., & Törk, K. (2021). *Caves and karst of Turkey* (Vol. 1). Springer AG.

2 Küre Mountains National Park

2.1 Regional Geography and Ecology

Küre Mountains National Park is located northwest of Turkey, within the provincial borders of Kastamonu and Bartın. The district centers around the national park are Azdavay, Pınarbaşı, Ulus, Kurucaşile, Amasra, and Cide.

Unlike many other national parks, Küre Mountains National Park is characterized by its plateau-like nature, extending in an east-west direction and serving as a physical and social boundary for the surrounding areas. As a result, there are only a few settlements within the park's vicinity. The buffer zone is the area immediately surrounding the national park, and the planning area includes the buffer zone. The national park covers a total area of 370 km^2, while the buffer zone covers approximately 800 km^2 (Fig. 2.1). Although there are about 50 villages in the buffer zone, the total population living in this area is less than 10,000 due to migration to larger cities. Nevertheless, small towns such as Azdavay, Pınarbaşı, and Cide, despite their distance from the buffer zone, maintain an organic relationship with the national park.

The inception of the Küre Mountains National Park began in the 1980s when engineers conducting forest management plans in the region first discovered its awe-inspiring natural features. Survey studies commenced in 1995, and the project "National Parks and Protected Areas" initiated in 1997, involved collaboration with the United Nations Development Program (UNDP), Food and Agriculture Organization of the United Nations (FAO), and the Ministry of Forestry, General Directorate of National Parks. Recognized as one of the priority areas for the protection of European forests by the World Wildlife Fund (WWF) in 1998, Küre Mountains became a national park by the decision of the Council of Ministers in July 2000. The designation of this region as a national park was Turkey's contribution to the United Nations' declaration of the "2002 World Year of Mountains."

With the research carried out in the project "Improving Forest and Protected Area Management" supported by the Global Environment Facility (GEF), Küre Mountains National Park received the PAN Parks certificate in April 2012, which symbolizes Europe's elite protected areas, becoming Turkey's first and Europe's 13th PAN Parks certified area. The goal of PAN (Protected Area Network) Parks is to create a network of natural sites in Europe that preserve their wild originality and minimize the human impact.

The defining feature of Küre Mountains National Park consists of longitudinally oriented high ridges and the deep valleys between them, covered with dense forest vegetation. The ridges predominantly comprise limestone, while the valleys consist of marl or flysch. These landforms and lithologies will be further described in subsequent sections. The region's abundant rainfall contributes to the dense forest cover, which, in turn, enhances its biological diversity. The area outside the national park, designated as the buffer zone, exhibits a more developed landscape at lower elevations compared to the national park itself. Several elevations can be found in the region, including Armutçalı Hill (1443 m), Kezboğazı (1299 m), and Üçgöknar Hill (1300 m).

The most important streams in Küre Mountains National Park and its buffer zone are Arıt, Ulus, Drahna, Aydos, and Devrekani. Arıt Stream is located in Bartın Province and in the buffer zone of the National Park, and its drainage length is 137 km. Ulus Stream, which drains the Ulus Basin, has a very high flow rate. It flows in an east-west direction and joins the Drahna Stream to form the Ulukaya Gorge. The Aydos Stream, which flows in a wide arc and has a length of 60 km, flows into the Black Sea after passing through the Aydos Canyon.

The largest stream in the Küre Mountains National Park and Buffer Zone is the Devrekani Stream. It begins in Devrekani and flows north along the Valla Canyon. The main course of the Devrekani, which is fed by two separate streams at the beginning of the canyon, is 147 km long and has a high season flow of 40 m^3/sec (Figs. 2.2 and 2.3).

All these streams have formed impressive canyons in the national park and buffer zone. Important examples are the

A. Yamaç, E. Tok, *Caves and Karst of Küre Mountains National Park, Turkey*, Cave and Karst Systems of the World,
https://doi.org/10.1007/978-3-031-80142-6_2

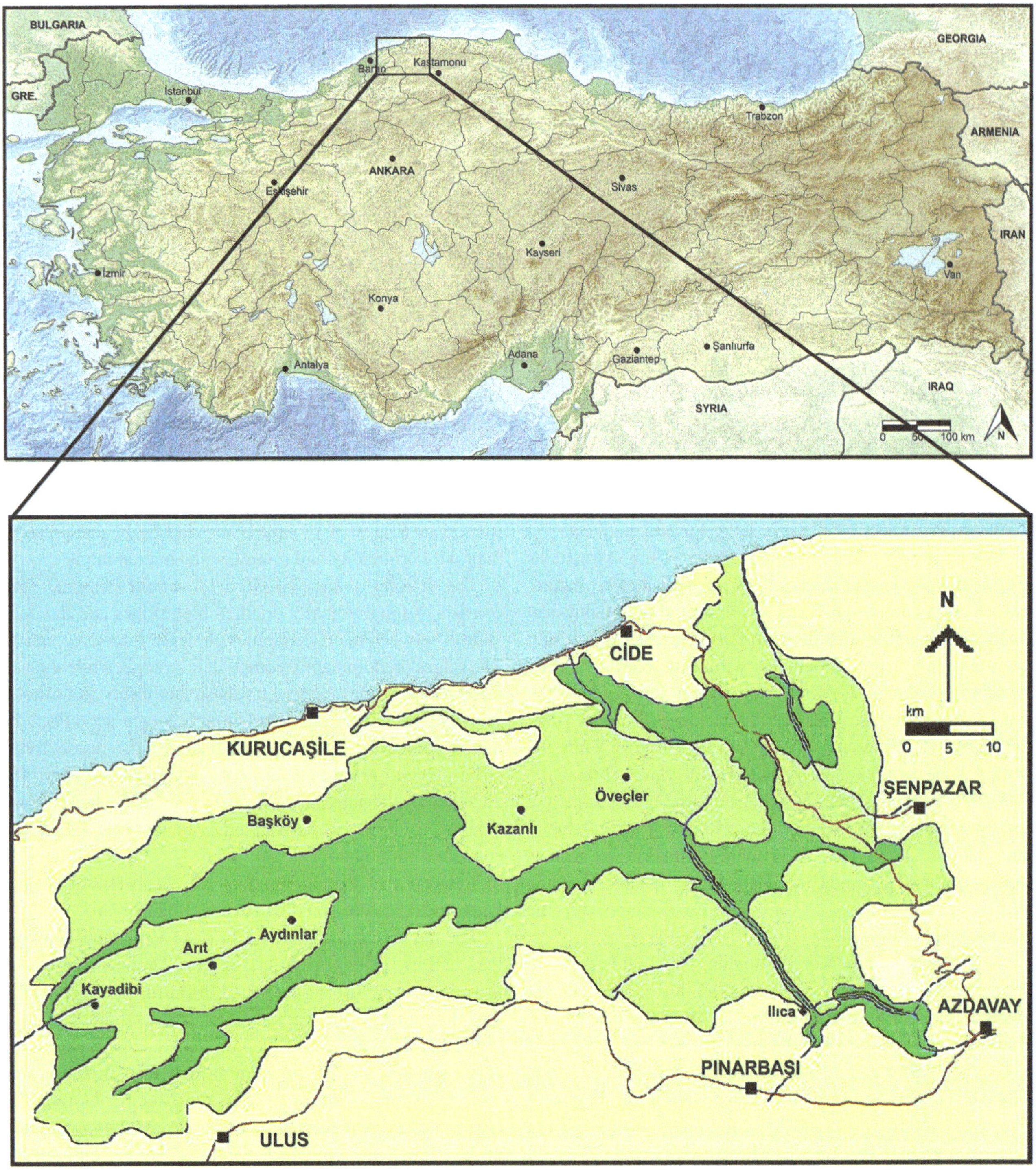

Fig. 2.1 Location map showing the borders of Küre Mountains National Park and the buffer zone (after Google Maps-elaboration A. Yamaç)

Valla, Ulukaya, Aydos, Loç, Çatak, and Horma canyons. The most important of these is Valla Canyon, which is 10 km long and up to 800 m deep in places, making it the largest and most impressive canyon not only in the region but in the whole of Anatolia (Güney, 2021).

Unfortunately, the region's streams, characterized by high flow rates, are often subject to tempting hydroelectric power plant projects. While the entire region should be recognized and protected as a natural wealth, the fact that it is seen as a source of income attracts private companies trying to implement these projects in the region. The preservation of the

Fig. 2.2 A branch of the Devrekani Stream with an old bridge, Yaylaköy—Armutlu road (photo A. Yamaç)

natural formations in the region faces significant threats due to these ongoing debates and efforts.

One of the most striking features of the Küre Mountains is its towering karst plateau and colossal canyons. The plateau, which forms the core of the national park, overlooks the surrounding areas and is enclosed by karst walls that are hundreds of meters high. Limited access to the plateau has helped to protect wildlife from outside influences. In addition to its geological and hydrological importance, the Küre Mountains National Park boasts a rich biodiversity, both in terms of flora and fauna.

Küre Mountains National Park and its buffer zone, which is an "Important Nature Area," "Important Bird Area," and "Important Plant Area" according to international protection criteria, is home to 930 plant taxa. Of these, 157 plants are endemic and 60 taxa are endangered. The region is home to 48 of the more than 160 mammal species living in Turkey, including mammal species such as the wildcat (*Felis sylvestris*), otter (*Lutra lutra*), brown bear (*Ursus arctos*), and red deer (*Cervus elaphus*). The small vultures (*Neophron percnopterus*) living in the national park, where 129 bird species have been identified so far, are globally endangered. In addition, 113 invertebrate species, 10 amphibian species, and 23 reptile species have been identified to live in the Küre Mountains National Park (Tekebaş, 2017).

On the other hand, it has been documented by various studies that the current flora of the region has been preserved unchanged for hundreds of years. For example, according to the results of archaeobotanical research conducted by Akkemik in Ilgarini Cave, the vegetation around the region 1000 years ago was cold high mountain forests, composed of mainly broad-leaved forest trees such as *Fagus orientalis* Lipsky., *Corylus colurna L., Carpinus betulus L., Carpinus orientalis Miller, Ostrya carpinifolia Scop., Castanea sativa*

Fig. 2.3 Valla Canyon (photo A. Yamaç)

Miller, Quercus L. sp. at 700–1000 m a.s.l., and *Abies bornmülleriana Mattf.* The area at present has the same forests. In addition to these species, *Pinus nigra Arn* and *Pinus sylvestris L.* are located in the area. The climate type of the area was humid and mesothermal, and there was a middle-level water deficiency in the summer (Akkemik et al., 2004).

The biological and geological richness of Küre Mountains National Park are intimately interconnected. The ecological systems are closely tied to the karstic geomorphology present in the park. The exceptionally well-preserved karst ecosystems within the national park are globally rare and of significant importance for nature conservation.

Küre Mountains National Park and its buffer zone are home to the best-preserved examples of the endangered "Black Sea Moist Karst Forest" ecosystems. These forests are considered one of the 100 Forest Hotspots in Europe, deserving special protection. Moreover, the park falls within the "North Anatolian and Caucasus Temperate Zone Forests," one of the 200 globally prioritized ecological regions selected by the WWF for nature conservation (Dokay., 2018; Tekebaş, 2017).

The park is within the Euro-Siberian Floristic Zone's Euxine part in terms of plant geography. The region's vegetation can be categorized into three distinct groups:

- Western Black Sea temperate zone beech and fir forests
- Pseudo-shrub formations
- Karst area mixed forests

The core area of Küre Mountains National Park, enclosed by cliffs and canyons, hosts natural, ancient mixed-leaved forests (including beech, hornbeam, chestnut, maple, and ash) and coniferous forests (such as the endemic Uludağ fir, larch, and Scotch pine). The region exhibits a homogeneous forest characteristic, with a combination of deciduous and evergreen tree and shrub species prevalent in temperate zones. The core area, specifically, showcases a vegetation structure indicative of humid conditions due to the dominance of both deciduous and evergreen species. The lower areas near the Black Sea coast display Mediterranean elements and shrubs. The region's significance lies in the diversity of its ecosystems and habitats, including natural forests, degraded forests, traditional agricultural ecosystems, and river ecosystems. Additionally, micro-ecosystems such as canyon ecosystems and forest meadow ecosystems contribute to the region's ecological diversity.

The western part of Küre Mountains National Park showcases distinct forest compositions owing to its karst terrain and humid atmospheric conditions. The unfragmented ancient forest ecosystems on steep karst slopes remain well preserved. These environments, particularly the karst depressions and desolate valleys south of Cide, support unique tree and shrub species associated with moist forests. Some notable examples include boxwood (*Buxus sempervirens*), hop-hornbeam (*Ostrya carpinifolia*), mountain ash (*Sorbus aucuparia*), Turkish hazelnut (*Corylus colurna*), hornbeam (*Carpinus betulus*), elderberry/dane weed (*Sambucus ebulus*), Norway maple (*Acer platanoides*), and holly (*Ilex aquifolium*) (Tunçkol et al., 2022) (Fig. 2.4).

Fig. 2.4 On the route to Ejder Cave, through a dense forest (photo O. Chachula)

Considering the density of vegetation in the region, its positive effect on karstification will be understood. In regions with substantial precipitation, the presence of CO^2 in rainwater is significant. Despite the thickness of the soil cover, the high amount of water enables the penetration of CO^2 into the soil, thereby increasing chemical dissolution. This situation, known as "covered karst," is very effective in the region.

The karst depressions on the high plateau create relatively damp corners in which specimens of giant trees can still be seen. In addition to the trees listed above, there are the red-twigged lime (*Tilia rubra*) with a height of 35–40 m and a trunk diameter of 80–90 cm, ash (*Fraxinus angustifolia*), oriental beech (*Fagus orientalis lipsky*), and wych elm (*Ulmus glabra*) (Güney, 2021; Güney et al., 2012).

Küre Mountains National Park stands out as a significant area in Turkey in terms of endemic plants. Recent studies have identified 250 endemic plant taxa within the park's boundaries, contributing to the knowledge and expanding the list of endemic plants of Anatolia (Tunçkol et al., 2020, 2022). Given the park's location within the transitional zone between the Euro-Siberian and Irano-Turanian flora regions, the distribution of endemic flora is influenced by these two regions. However, more endemic taxa are present in the Euro-Siberian region. Recent research in the Küre Mountains National Park led to the discovery of additional new species. *Polystichum asiae-minoris* was one of them (Fig. 2.5) (Tunçkol et al., 2020).

The endemic and new species discovered during the local flora studies indicate that a conservation status for these species should be given. In the area where *Polystichum asiae-minoris* was discovered, a new walking trail has put the population in danger. Therefore, it is crucial to stop any activities that endanger the plants and their natural habitats in the area. Additionally, steps should be taken to stop activities such as road construction, marble quarries, and deforestation.

The area's climate can be characterized as a transitional climate between the Black Sea and the central Anatolian continental climates, with both maritime and continental influences. The Küre Mountains, running parallel to the coast, act as a barrier, gradually diminishing the influence of the Black Sea climate toward the interior while introducing the harsher and more continental characteristics of Central Anatolia's climate. The buffer zone experiences an overall temperate climate, with warm and humid summers and cold, snowy winters. Average temperatures during the summer months range from 20–25 °C, while winter temperatures can drop below freezing with frequent snowfall. Although snowfall in low-lying areas may not persist continuously, it lingers until mid-spring at higher elevations. The region receives ample precipitation throughout the year, primarily influenced by the Black Sea. Rainfall is particularly abundant during autumn and winter, with an annual average ranging between 800 and 1000 millimeters. These precipitation patterns nour-

Fig. 2.5 Some endemic plants of Küre Mountains National Park; 1. P. Asiae-minoris, 2. V. spectabile var. isandrum, 3. C. Wendelboi subsp. cognesta, 4. P. Graminifolium (Tunçkol et al., 2020)

ish the region's forests and rivers, contributing to its natural beauty. The prevailing winds in the region originate from the west and northwest, bringing coolness and humidity from the Black Sea. During the winter months, these winds can be quite strong.

The table below presents the 30-year average temperatures for the Pınarbaşı district (°C):

Months	1	2	3	4	5	6	7	8	9	10	11	12
Average	−1.6	−0.3	3.5	8.8	11.7	15.7	18.1	18.6	14.0	9.8	4.7	0.7
Highest	14.2	20.0	25.0	30.0	33.1	34.2	36.8	36.6	35.4	31.0	24.3	19.0
Lowest	−24.0	−24.3	−15.7	−6.1	−4.8	2.0	4.2	3.8	−2.0	−4.2	−20.7	−19.3

Fig. 2.6 A winter camp near Pınarbaşı (photo Ç. Çankırılı)

The climatological data for the area were obtained from the Directorate of State Meteorology. These data are Pınarbaşı district values and belong to measurements made between 1985 and 2015. It should be noted that the National Park is approximately 600 m higher than Pınarbaşı district, which is at an altitude of 650 m (Güney, 2021) (Figs. 2.6 and 2.7).

Fig. 2.7 Crossing the Devrekani River towards Gökdeli Cave during winter (photo Ç. Çankırılı)

2.2 Regional Geology

Forty kilometers east of Bartın lies one of the Küre Mountains National Park's most breathtaking views. Here, just north of Söğütlü Village, Gürdek Rock rises to the sky like a white wall (Fig. 2.8).

This particular area of the national park preserves the remnants of crustaceans that settled on the seabed around 130–150 million years ago, during the Late Jurassic to Early Cretaceous periods. Over an extensive period, these crustaceans underwent a transformation into limestone through the slow process of sedimentation and subsequent consolidation of the calcite mineral present in their remains under intense pressure. As a result, these limestone formations eventually emerged from beneath the Earth's surface, forming stacks that are hundreds of meters thick.

The formation of limestone is a remarkable phenomenon in nature. When crustaceans from marine or freshwater environments perish, their remains sink to the bottom of the water, accumulating over an extended period. This accumulation leads to the formation of sedimentary layers composed of millions of such crustaceans. Over time, the calcite mineral present in their bodies gradually solidifies, resulting in the formation of thickened layers. Eventually, the original crustaceans disappear, leaving behind rocks comprised of their mineralized remains. This geological process occurs at such a gradual pace that it eludes direct observation. In fact, it takes approximately 10,000–12,000 years for the lime-

Fig. 2.8 Gürdek Rock Cliff (photo O. Chachula)

stone formed by the sedimentation of these crustaceans to reach a height of just one meter (AMC Şengör, 2000).

The concept of time in geology poses a significant challenge for human comprehension. Our imaginations are confined within the span of our own lives, making it difficult to envision time scales beyond that. To provide a classic analogy often described, if the approximately 4 billion years of Earth's existence were compressed into a 24-h day, the emergence of humanity within the last 2 million years would represent merely the final minute of that day. In contrast, geology encompasses the remaining 23 h and 59 min (Fig. 2.9).

Similar to how limestone forms, many other phenomena are difficult to perceive due to this lengthy geological process. For example, the Earth is a dynamic structure, even though we might not observe it in our brief lifetimes.

The theory that continents move was first proposed by Alfred Lothar Wegener in 1912 and was ridiculed, let alone accepted, by the scientific world. However, he had analyzed both sides of the Atlantic Ocean for rock type, geological structures, and fossils and noticed many similarities. This concept, called "continental drift" by Wegener, formed the basis of today's plate tectonics. This theory states that the lithosphere of Earth is made up of numerous minor plates in addition to eight major tectonic plates, all of which have been slowly moving for approximately 3.4 billion years. Where the plates meet, their relative motion determines the type of plate boundary (or fault).

The North Anatolian Fault is one of the longest boundaries of these minor plates. Between the Eurasian Plate and Anatolian Plate transform boundary, this fault line crosses whole North Anatolia and is 1200–1300 km long. It has a right-lateral strike-slip movement, indicating that the south block of the fault continuously shifts westward. This tectonic movement occurs at a rate of approximately 2.2 cm per year (AMC Şengör et al., 2005). Over the course of 100 years, the accumulated strain along this fault is released through earthquakes, resulting in a horizontal displacement of 2.2 m. Furthermore, when we consider the cumulative displacement of 2.2 cm each year, it becomes nearly impossible to fathom the future position of Anatolia in 100 million years.

Anatolia presents an unfortunate aspect in its geological history. While regions such as England, Siberia, West America, and Scandinavia have remained relatively cohesive landmasses for millions of years, Anatolia has experienced constant fragmentation, dispersal, reassembly, uplift, and subsidence throughout this extensive timeframe.

If we delve further back in time, to when life first emerged on Earth, all the continents were once joined together as a single landmass known as Pangea. Approximately 400 million years ago, Pangea began to split into two, resulting in the formation of the southern continent named Gondwana and the northern continent named Laurasia.

Around 255 million years ago, during the Early Triassic period, these two immense continents existed, with the Paleo-Tethys Ocean occupying the expanse between them. Anatolia became part of North Gondwana during this period, while some portions remained submerged beneath the Paleo-Tethys Ocean (Fig. 2.10).

In recent studies, trace fossils of *Hyloidichnus* footprints dating back to the Cisuralian period (270–300 million years ago) were discovered in the Çakraz Formation of northwestern Anatolia. Additionally, plant fossils such as *Annularia*, *Stigmaria*, and the *Coniferophyte walchia* were unearthed from the same region during the survey. These fossils, along with sedimentological evidence such as mud cracks and raindrop imprints, indicate the presence of captorhinid reptiles that inhabited palustrine floodplains characterized by warm, seasonal climates alternating between humid and relatively prolonged dry periods (Gand et al., 2011).

Rifting refers to the process by which the Earth's hard crust splits apart due to the expansion of the crust caused by heat emanating from the Earth's core. In Anatolia, the surface portions gradually fractured as a result of this rifting. According to stratigraphic and sedimentological data, rifting in the northern region of Gondwana began earlier. However,

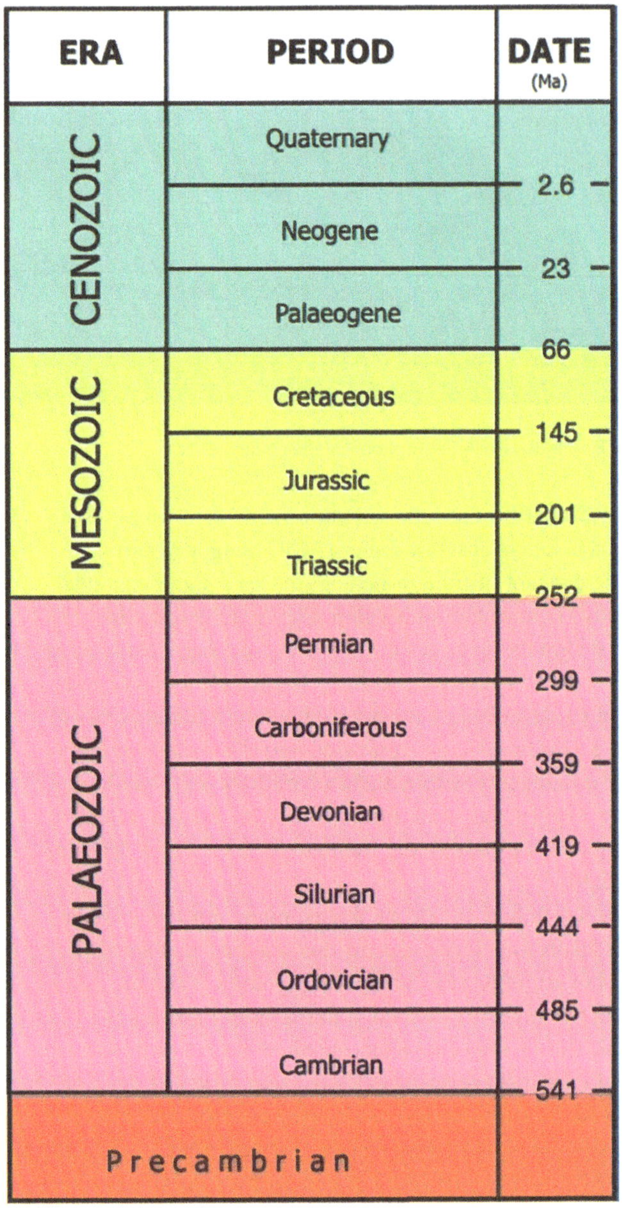

Fig. 2.9 Geological time scale (after Wikipedia-elaboration A. Yamaç)

the separation of the Anatolian microcontinent from Africa commenced around 205 million years ago, marking the end of the Triassic period. This delay in separation is attributed to the thinning of the crust that occurs between the onset of rifting and the final detachment.

Approximately 190–200 million years ago (Early Jurassic Period), as the Laurasia and Gondwana plates converged, ultimately closing the Paleo-Tethys Ocean between them, the Anatolian landmass, which still contained marine sediment at the time, experienced a southwest-to-northeast movement resulting from tectonic forces. This period also witnessed the emergence of facies from the Paleo-Tethys Ocean, likely along the southern coast. These shallow marine sediments can still be traced in regions of northwest and central Anatolia.

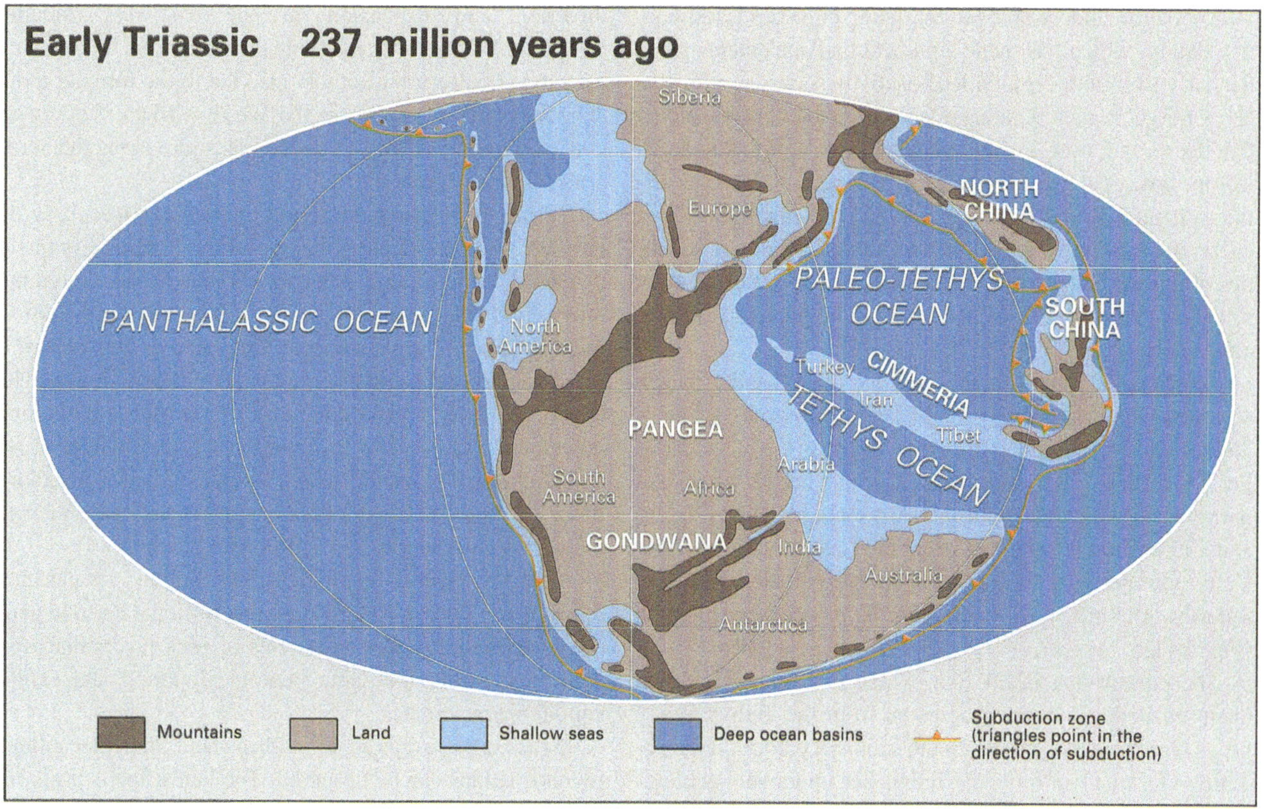

Fig. 2.10 World during Early Triassic (237 million years ago) (Paleomap Project)

A widely accepted hypothesis posits that during the closure of the Paleo-Tethys Ocean or the collision of the Laurasia and Gondwana continents, a thin continental plate detached from Gondwana and amalgamated with Laurasia during the Paleotethys' extinction phase. This tectonic event is believed to have occurred approximately 190 million years ago and influenced adjacent areas of Anatolia.

Now you're wondering how we can know with such certainty all these events that happened millions of years ago. The precise understanding of events that transpired millions of years ago in geology relies on centuries of dedicated research, the rectification of mistakes, and the progression of scientific knowledge. Geology encompasses various sub-disciplines such as geomorphology, petrology, and stratigraphy, which explore Earth's formation, geological time periods, and continental movements.

The science of paleontology, which studies fossils, played an essential role in determining the ages of geological formations even before the sixteenth century. However, it was only after 1840, with the observation of distinct fossils within different rock layers, that the classification of these fossils and the establishment of a standardized framework gained momentum. Paleontology has since developed into a global discipline, with millions of fossils classified based on their age, species, and distinguishing characteristics. Consequently, the discovery of a fossil within a rock provides valuable insights into the age of that rock. Index fossils, also known as guide fossils or indicator fossils, are especially crucial for age determination. These fossils serve as markers for defining and identifying geologic periods or faunal stages. For example, the presence of fossils like *Beneckeia buchi* or *Encrinus liliformis* in a rock confirms its belonging to the Middle Triassic Period, dating back approximately 228–245 million years ago.

Another method employed for dating purposes is radioactive dating, which has made significant advancements since the early twentieth century. Radioactivity serves as a precise clock for geologists. Radioactive isotopes are unstable atoms that decay over time, eventually reaching a stable state. Each isotope decays at its own rate, such as Uranium-235 converting to Lead-207 at a rate where half of the atoms transform every 704 million years. By measuring the types and quantities of individual radioactive atoms present in a rock sample, scientists can calculate its age through the process of radioactive decay.

Returning to Anatolia, the land we inhabit today began separating approximately 190 million years ago as the Paleo-Tethys Ocean continued to close. During the Cretaceous Period, around 150 million years ago, this ocean, which had narrowed significantly, split into two sections. Between 130 and 150 million years ago, during the Late Jurassic to Early Cretaceous periods, the Anatolian microcontinent underwent a northward shift, positioning it in its present-day location (Fig. 2.11).

The present-day Black Sea, which is a relatively small enclosed body of water, originated from the Tethys Sea, a much larger sea that existed in the southern part of the Paleo-Tethys Ocean. Over time, the Tethys Sea underwent geological processes such as faulting, rupture, block transformation, and thrusting, resulting in its contraction in the north-south direction and expansion in the east-west direction 90–110 million years ago (during the Aptian and Cenomanian periods). This tectonic activity also led to the formation of a significant magmatic arc in the Western Black Sea region, characterized by numerous active volcanoes along the northern coastline of Anatolia.

During this geological period, Anatolia, particularly the northern continental block known as the Pontide Continent, exhibited an intriguing behavior. While it was commonly understood that rifting typically led to the formation of microcontinents, Anatolia first fragmented into relatively small islands that could not even be classified as microcontinents and later reassembled due to subsequent tectonic movements. In contrast to regions such as England or the Scandinavian Peninsula, which maintained their structural integrity, Anatolia experienced a continuous cycle of fragmentation, integration, and subsequent breakup.

After this general geological information, we can now return to our region: Küre Mountains National Park. In geology, a formation refers to a series of rock layers that were deposited during the same geological period and exhibit similar characteristics.

In the northern region of Pınarbaşı, near Sümenler Village, two formations can be observed. The landscape is primarily covered by dense vegetation, with few discernible rock outcrops, as one walks through a breathtaking forest that is

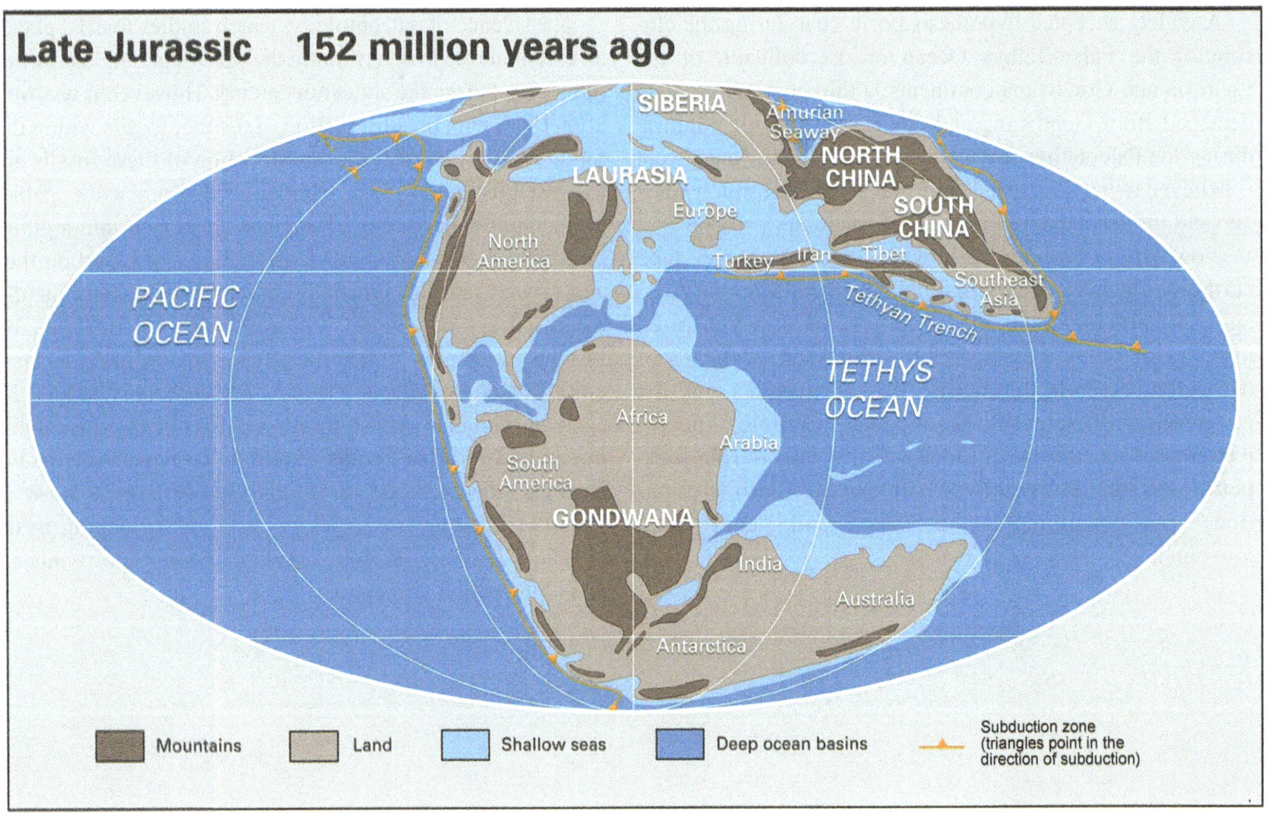

Fig. 2.11 World during Late Jurassic (152 million years ago), some fragments of Turkey and Iran were on the border of Tethyan Trench (Paleomap Project)

dominated by centuries-old trees. However, further north, near Topmeydanı, the forest recedes, revealing a limestone wall. At this point, the Ulus Formation, which has been traversed thus far, comes to an end, and another formation, known as the İnaltı Formation, begins.

Ulus Formation, which takes its name from Bartın—Ulus Village and is widely observed on the east—west line from there to Sümenler Village, consists of various rock types such as sandstone, claystone, shale, and turbiditic rocks. It contains macrofossils like *Ancyloceras* sp., *Deshayesites* sp., and *Chelonoceras* sp. (Fig. 2.12).

Since it is turbiditic, frequent facies and thickness changes are observed. Numerous gravitational shifts observed in the turbidites indicate that tectonic movements continued uninterruptedly both in the basin character and during sedimentation. The presence of igneous and volcanic inputs shows the mobility of the basin calcareous megabreccia and volcanite blocks are also observed along the tectonic zones. The age of Ulus Formation is determined as Early Cretaceous (Derman, 1995).

The rocks overlying the Ulus Formation and forming the entire plateau of the Küre Mountains National Park are the İnaltı Formation. Formed and deposited 130–150 million years ago, between the Late Jurassic and Early Cretaceous periods, these rocks consist entirely of limestone. The thickness of this formation is estimated to be 800–1000 m. It consists of white-gray colored, medium-thick layered, neritic-shelf limestones with abundant fossils in places (Gedik & Aksay, 2002; Kazancı et al., 2009) (Fig. 2.13). Although it is named after the village of İnaltı in the south of Sinop, the horizontal and vertical facies changes are best seen in Yarpınar in the south of Cide. This geological formation, which covers large areas in the Küre Mountains National Park, starts from the north and south of Arıt and surfaces in a general east-west direction towards northeast towards Şenpazar. İnaltı limestones show a steep topography in the area. High slopes, ridges, and karstic structures are the most important karstic and morphological features. Karstification in İnaltı limestones is developed especially along the tectonic lines (Tüysüz, 2022).

130 million years ago, when these rock formations were accumulating, the oceans were inhabited by marine reptiles like *Ichthyosaurs* and *Plesiosaurs*, while the land was dominated by dinosaurs such as *Camarasaurus, Apatosaurus, Diplodocus,* and *Brachiosaurus.* These terrestrial giants coexisted with predators like *Ceratosaurus, Megalosaurus, Torvosaurus,* and *Allosaurus.* Simultaneously, birds were beginning to emerge, and the skies were ruled by *Pterosaurs.*

North of the villages of Sümenler and Kazla, the Ulus Formation sank beneath the Inalti Formation 84 million years ago and uplifted the latter, converting it into a plateau. This geological event is visible in the subduction zone that forms the southern boundary of the Küre Mountains National Park (Aylar et al., 2018). Since the top of this plateau does not allow any settlement due to its steep walls, it has become naturally protected and home to wild nature. So, in a sense, this geological event ultimately shaped Küre Mountains National Park into its current protected and wild nature-filled state (Fig. 2.14).

This zone is defined by the impermeable Ulus Formation subducting beneath the İnaltı Formation. Consequently, streams flowing from south to north on the impermeable formation also sink below the İnaltı Formation, resulting in the formation of numerous caves, with sinkholes being the most common type. Approximately one-fifth of the caves in the Küre Mountains National Park are located in this subduction zone. The positioning of the caves in this region can be seen in Chap. 3, especially in Fig. 3.46.

This impermeable layer, the Ulus Formation intersecting with the Valla Canyon in the east, extends along the western border of the national park, encompassing the Sümenler-Sorkun-Dizdarlı-Yukarıdere-Alıçlı line. Active sinkholes are found wherever this impermeable layer interacts with the underlying limestone block. The majority of these sinking waters are believed to discharge into the Valla Canyon. Additionally, it is inevitable that some other aquifer caves, such as the Ulukaya Cave along the same line, drain the water from the sinkholes that have formed on the plateau.

The plateau's highest point is occupied by the İnaltı Formation, while the northern and Arıt regions are characterized by the Upper Jurassic Çakraz Formation. It is worth noting that the geological map of Küre Mountains National Park closely aligns with the distribution of formations, with the Çakraz Formation almost entirely lacking limestone, while the İnaltı Formation dominates the plateau's summit.

Fig. 2.12 *Ancyloceras* sp. (left) and *Deshayesites* sp. (middle and right)

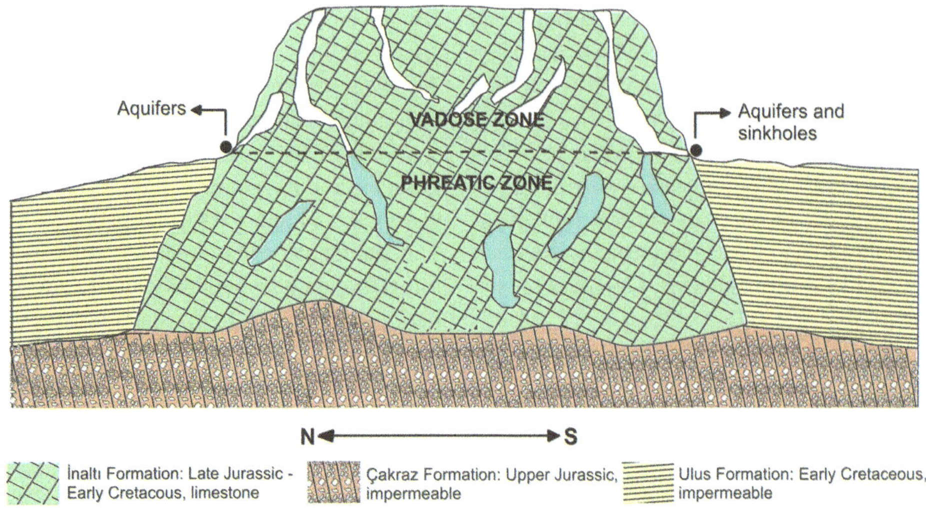

Fig. 2.13 Schematic cross-section of three different rock formations around Küre Mountains (drawn A. Yamaç)

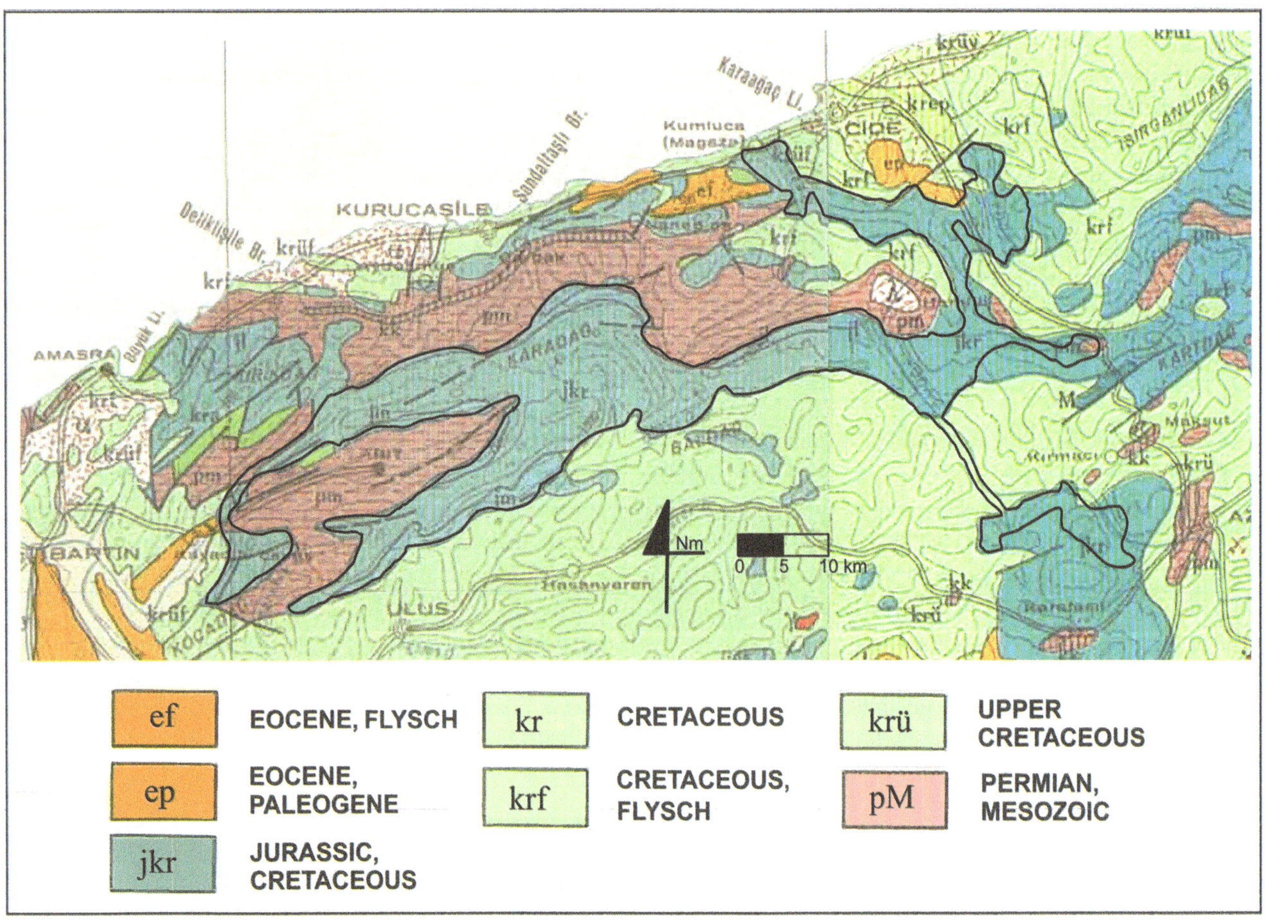

Fig. 2.14 Geological map of the region and Küre Mountains National Park (MTA 1964-elaboration A. Yamaç)

The National Park area, which extends in the east-west direction parallel to the Black Sea coast, has a morphologically high plateau appearance, while the Buffer Zone consists of sloping foothill plains (pediment) at lower levels than the National Park. The area, which is generally formed by medium-height mountains, is divided by deep valleys in places. The main geomorphologic structure in the research area located between Kastamonu—Bartın settlement areas and the Black Sea is characterized by elevations and declines. Block tectonism has an important effect on the morphology of the region. The effects of lithologic and paleoclimatic features cannot be ignored, along with the tectonic activities that seem to be intense. The three main morphologic structures, namely, the descent-elevation and re-decline, extending roughly parallel to each other from south to north, are very prominent (Güney et al., 2012).

Millions more years pass after the meeting of these two formations. As a result of continental movements, the Tethys Ocean has shrunk and turned into a sea. This sea will form the Mediterranean, Black Sea, Caspian, and Aral lakes in the future. In this process, the continent of Gondwana was also fragmented and divided into many small continents. One of the continents formed by the disintegration of Gondwana was Cimmeria. We know it is hard to believe that the regions that are Anatolia, Iraq, and even Tibet today belonged to this continent 65 million years ago (Okay, 2008).

The Late Cretaceous period saw the emergence of fascinating fossils, including a skull fragment of *Mosasaurus hoffmanii*, a giant marine creature about 17 meters long that lived in the oceans during the time of the dinosaurs. This significant fossil was found in 1999 at the construction site of the Beyler Dam, north of Devrekani, by Hacettepe University's Cemal Tunoğlu and introduced to the scientific world (Tunoğlu & Bardet, 2006) (Fig. 2.15). Additionally, a new foraminiferal genus named Cideina, also dated to the Late Cretaceous (Maastrichtian), was found near Cide (Sirel, 1991).

The Late Cretaceous period was marked by the appearance of the first mammals as well as significant tectonic movements that created large-scale magmatism and orogeny (mountain formation), such as the Balkanite, Pontide, Crimean, and Caucasian uplifts. These tectonic movements were influenced by the ongoing collision between the Eurasian and African plates and the westward escape of the Anatolian block. The North Anatolian Fault, which is now a prominent feature in the region, would form a few million years later in the Late Miocene (approximately 7 million years ago) as a result of these tectonic processes.

In the Cretaceous, east-west-oriented horsts and grabens began to form due to tectonic events, and the basin collapsed slowly at first and then rapidly during this period. The folds and fractures formed during this tectonic compression due to the push from south to north as a result of the collision of the continents in the Paleogene were the main factors in the region's present morphology. The region is still in a subduction phase and is located in an active earthquake zone. These faults were formed as a result of north-south or northwest-southeast-oriented compression and stresses. (Fig. 2.16).

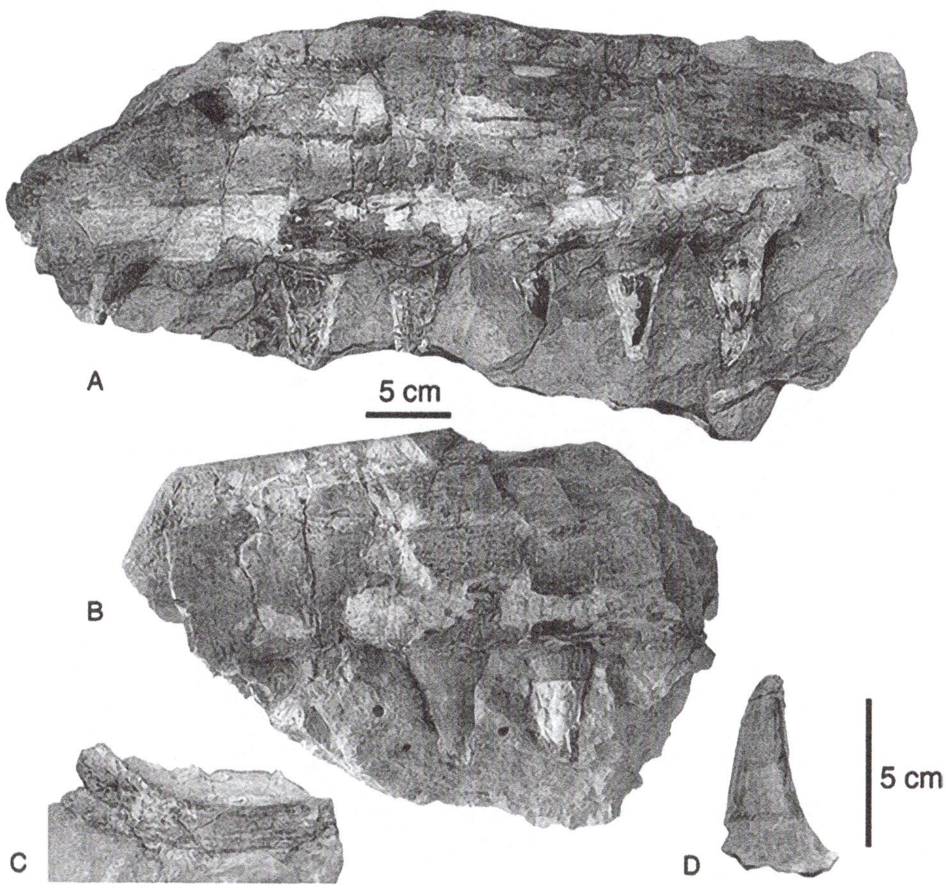

Fig. 2.15 Skull fragment of Mosasaurus *hoffmanii* (Tunoğlu & Bardet, 2006)

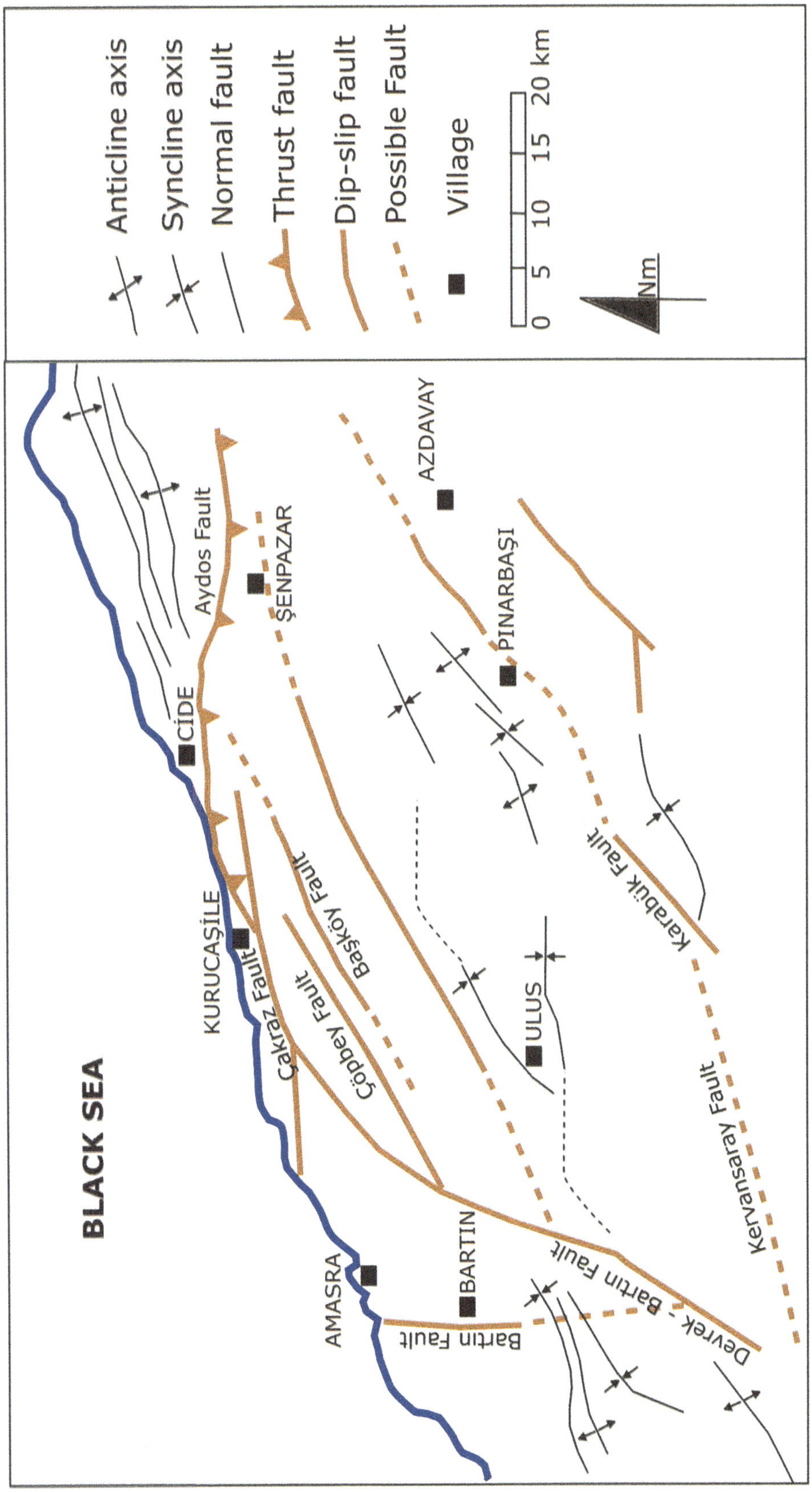

Fig. 2.16 Faults in the vicinity of Küre Mountains (Derman, 1990)

2.3 Regional Archaeology

It is well known that the area between the provinces of Kastamonu and Bartın, which includes the Küre Mountains National Park, has a history that dates back to the Paleolithic era.

However, the Black Sea, especially the Western Black Sea region, has for some reason been completely ignored from an archaeological point of view. It is difficult to understand the reasons behind this approach, because this region, like the rest of Anatolia, was inhabited continuously from the Paleolithic to the Classical Ages.

Surveys conducted in the region show that settlements began as early as the Paleolithic Age. In the 1940s and 1950s, Paleolithic tools were found in many different areas from Gölköy to Devrekani Mountains (Kökten, 1948; Şenyürek, 1944).

Following these pioneering studies, only two comprehensive surveys have been conducted in the region. The first of these was the Kastamonu Archaeological Survey. It began in 1995 and spanned three seasons, covering a sizable area. During this survey, more than a hundred findings, structures, and settlements dating from the Paleolithic to the Middle Ages—the majority of them previously undiscovered—were located and inventoried. On the other hand, unlike the Cide Archaeological Survey, which will start 14 years later, no caves were investigated during this survey. No research has been done in the area of Küre Mountains National Park within this survey, which was primarily concentrated in the east and southwest of Kastamonu city, and a total of six different reports have been published (Kuzucuoğlu et al., 1997; Marro et al., 1996, 1998; Özdoğan et al., 1996, 1997, 1998).

Since the 1950s, studies conducted in the Western Black Sea region have emphasized that few Late Bronze Age finds have been obtained or that the finds have gradually decreased towards the coast. During the surveys conducted by C. A. Burney, who directed one of these studies in the region, it was reported that a small number of ceramics dating to the Late Bronze Age were recovered in the vicinity of Karabük-Eflani (Burney, 1956). In addition, A. Ünal emphasizes that Salman Höyük, located by the Devrez Stream, has an important potential for understanding the Late Bronze Age cultures of the Western Black Sea Region (Ünal, 1999). Semercitepe (Eflani) and Taşköprü Yolu mounds are also thought to be among the important Late Bronze Age settlements of the region. As we move from the interior of the region towards the coast, it is noteworthy that the number of Late Bronze Age mounds and sites in the mountainous and densely vegetated geography starting from the Karabük line gradually decreases. As mentioned above, the main reason for this situation is the difficult geographical conditions of the region.

In the caves and rock shelters identified during the Cide-Şenpazar Survey, which is one of the comprehensive surveys conducted in the Western Black Sea Region, a limited number of Late Bronze Age finds were recovered. Again in this region, a sword dating to the Late Bronze Age and enabling us to establish a direct connection with the Hittite culture was found in the Buz Cave in Pınarbaşı, Kastamonu (Ünal, 1999). Gökdeli (Spear—Buz) Cave, described in Chap. 3, Sect. 3.3.16, is the cave where this sword was found.

The surveys conducted by Matthews and his team in the area between Çankırı, Kastamonu, and Karabük provinces also yielded some finds that can be dated to the Late Bronze Age (Glatz & Matthews, 2005). In addition to the limited data obtained from the surveys conducted in the region, an important assemblage of finds was recovered in Kastamonu-Kınık (Emre & Çınaroğlu, 1993; Yıldırım et al., 2021). However, no settlement contemporary with the finds was identified during the excavations at the site of Delibeyoğlu Ridge.

Between 2009 and 2011, the Cide Archaeological Project was realized. During this project, under the direction of Düring and Glatz, a large area was surveyed in the coastal and mountainous parts of Cide (Fig. 2.17). Seven caves (Ballıcı Cave, Çadır Cave, Derebağ Village Cave, Gideros Cave, Kılıçlı Cave, Okçular Cave, and Uçakdibi Cave) have also been extensively surveyed as part of the same project. In some of these caves, the details of which you will read in Chap. 3, Paleolithic tools were found. It is striking that, even though there was no excavation, and it was only a survey, these tools were found on the floors of the caves. In most of these caves, there were uninterrupted findings from the Paleolithic to the Bronze Age and into the Classical Periods. Several articles and a book have been published about this project (Düring et al., 2012; Düring & Glatz, 2015; Glatz et al., 2011).

Both the surveys and excavations in the region are very limited. In Chap. 3, you will see how many archaeological findings in the caves in the mountainous regions have not yet been scientifically investigated. For example, among the caves we explored, Atak, Ilgarini, and Medil 1 has an incredible concentration of archaeological findings, and these caves have never been archaeologically studied.

Only five archaeological excavations have so far been carried out in these two provinces: Bartın and Kastamonu, which have a total surface area of 16,000 km^2. Two of these excavations—Pompeiopolis and Hadrianopolis—are local rescue excavations dating back to the classical age. Both excavations were quite irregular and were conducted by the museum directorate from time to time.

Begun in 1984 and intermittent for several years, the excavations at Pompeiopolis were led by L. Summerer for a period between 2006 and 2016 (Summerer, 2017; Summerer & von Kienlin, 2013). Although no remains were visible on

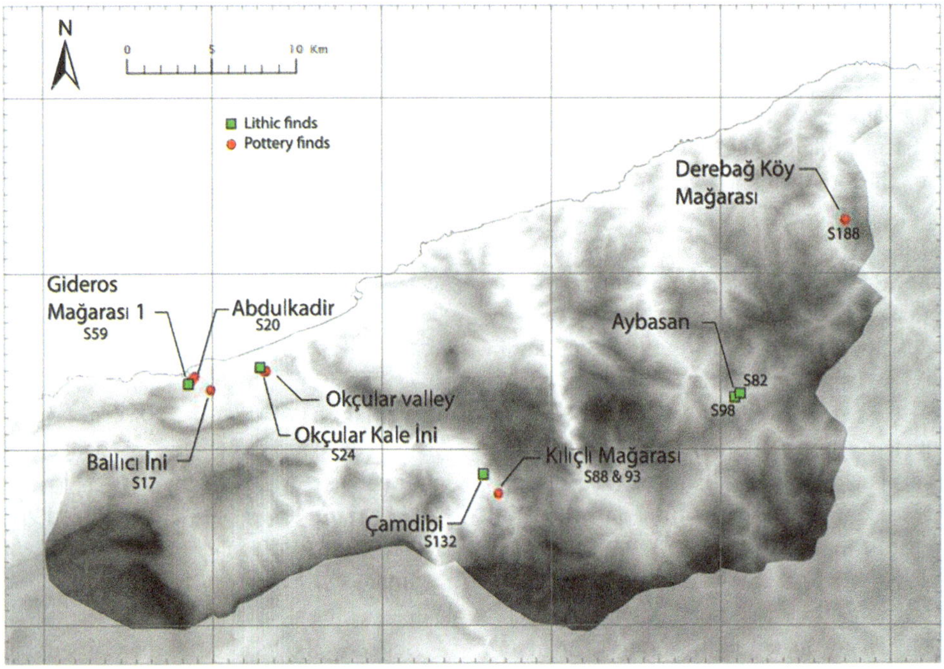

Fig. 2.17 Map showing the general survey area and some caves explored within the Cide Archaeological Project (Drawn T.C. Wilkinson) (Düring & Glatz, 2015)

the surface before the excavations, a systematic geophysical survey detected the outline of a large city with roads, public buildings, and two theaters that were successively archaeologically investigated. The excavations in the major theater unearthed the lowest marble seat rows as well as inscribed architraves with decorated friezes. From the inscription, it is understood that at least the stage was built around AD 150. The theater was dismantled within the fifth century. Scarce remains of an octagonal building were previously identified as the market building, while few Christian graves excavated in 2016 in this area have suggested a later use as a church.

Similarly, intermittent excavations at Hadrianopolis were carried out from time to time by the museum directorate and different archaeologists. The first scientific study on the city was the rescue excavation carried out by the Ereğli Museum in 2003.

Between 2005 and 2016, excavations were carried out by three different archaeologists.

Hadrianopolis, the most important Roman city in the Black Sea region, established in the first century BC and was occupied until the eighth century AD. It was named after the Roman emperor Hadrian in the second century AD. The city also had the names Caesarea and Proseilemmene.

It was situated in a rich agricultural area and controlled the principal western route from the Central Anatolian Plain through the mountains to Bartın and the Black Sea. Archaeological surface surveys have uncovered 14 public buildings and other structures in the ancient city. Among these public buildings are two baths, two churches, a defense structure, rock tombs, a theater, and other monumental buildings.

The third archaeological excavation in the region was the Selmanlı Tumulus Grave rescue excavation in 2012. In this rescue excavation, carried out in cooperation with Kastamonu Museum and Karabük University, many artifacts dating to the first century BC were found, although the tumulus had been previously looted by thieves. The 106 cm long painted high relief on the burial chamber wall is one of the most beautiful Late Hellenistic artifacts ever found in this region. The thieves were caught, and some of the objects they stole from the burial chamber were recovered (Ş. Yıldırım, 2013).

The remaining two archaeological excavations are Kahin Tepe and Devrekani-Kınık. Unfortunately, the Kahin Tepe excavations, which are important enough to shed light on the archaeological past of Northwestern Anatolia, continued for only three seasons.

For the time being, Kahin Tepe is the oldest settlement in the Black Sea Region of Turkey. Findings from the excavations here reveal that there were settlements dating back to the Early Bronze and Pre-Pottery Neolithic ages in the region (Fig. 2.18).

During the excavations of the site, votive pits and round structures dating back to 8000 years ago were unearthed. In addition, artifacts such as grinding stones and ornaments dating back to the Pre-Pottery Neolithic were also found. It is assumed that the animal figures and the figures on the obelisks are related to some kind of religious beliefs.

Fig. 2.18 Kahin Tepe excavation (photo N. Ayengin)

Two levels were reached before the excavations were halted in 2022 due to the dam construction in the region:

Layer I: Pre-Pottery Neolithic: Since no carbon trace was found in the excavation area, no clear dating could be made; however, based on the evaluation of the chipped stone industry in the area, as well as the architecture and artifacts presumably related to religious beliefs, it is believed that this layer dates to the Pre-Pottery Neolithic Age.

Layer II: Late Early Bronze Age to 2000 BC.

Even though Kahin Tepe produced an extensive number of findings during the brief excavation period, no scientific publication on it has yet been. Due to the short duration of the Kahin Tepe excavations and the lack of a comprehensive report, we do not have detailed information on the archaeological stratification (Ayengin, 2020).

Therefore, Devrekani-Kınık is the only excavation where we can observe the full stratification in the region. The excavations here yielded Late Chalcolithic/Early Bronze Age pottery finds from Layer I. The technical characteristics of the pottery from this settlement indicate that the settlement of Kınık has a similar culture with the Early Bronze Age settlements in Central Anatolia, especially in the Halys Region. Similar finds from the II. architectural phase of the settlement are frequently encountered in Early Hittite settlements. This situation shows us the Hittite cultural traces on the Paphlagonia Region in the second millennium BC (Genç, 2005).

2.4 History of Cave Explorations

After Prof. Cemal Arif Alagöz's first cave exploration in the region in 1944, no caving organization was interested in this area for about 40 years. The first speleological surveys in and around Küre Mountains were started by Boğaziçi University Cave Research Club (BÜMAK) in the early 1980s. Following these first studies, which you will read the details of below, the area quickly caught the interest of both domestic and foreign cavers, and research was accelerated.

In his work written in 1944, Prof. Cemal Arif Alagöz described this region as follows (Fig. 2.19) (Alagöz, 1944):

> Further east on the Black Sea coast, 4 km from Cide. To the southwest, it is passed in front of Devrekani Stream or Kocaçay. Within the white, hard-shaped limestones, a majestic V-shaped gorge draws attention. A karst land must be located behind these straits, you say. As a matter of fact, if we look at a detailed map, it can be seen that there are more than 15 covered pits, large and small, on the Kayaardı plateau to the north of Arıd, and to the east, between Bartın and Cide, and the Subati Stream to the west of Kayaboğazı.
>
> There are also straits and passages in this area: Kalem strait, Arıd's gate in the west, Çeme strait, and Döşeme pass... There are some names ending with a rock ('kaya' in Turkish) suffix: Arya Kaya, Tahtalık Kayası, Kuş Kaya, Sarı Kaya, Yanık Kayası Hill, Kayabaşı, Hacet Kayası, Kel Kaya, Efendiş Kayası, finally Kayaboğazı and Kokurdan plateau leave no doubt about the existence of a Bartın-Cide karst area. I made a short trip to the region (1937).
>
> In the vicinity of Kocaçay Straits, which is close to the Cide, marls are located at the bottom and limestone at the top. The gorge was excavated in limestone, but it didn't last long. All dark red masses separate the sloping limestone plateau in the south from the limestone of the gorges. With its layers starting from the Liassic to the Eocene, this place shows a folded structure and even a fish scale structure (Schuppenstruktur) in which the old layers cover the new ones. It is understood that the Kocaçay straits are located within the neocoma limestones. Naturally, karst areas will be discontinuous in a folded area where the limestone occurs in narrow or broad stripes. Lebling says the presence of depression lines in line with the fold and a string of doline parallel to the shore along these depression lines.
>
> We had just entered the Kocaçay Strait. Now let's move on. The limestone holes on the valley's walls, which look like they are lined up on a row at another height, are found among the green

Fig. 2.19 Cover of Alagöz's book

heaths. Since the waters of the stream cover the bottom of the valley where the valley narrows, it is almost necessary to perch on the horse while passing through these places. It is seen that some karst springs emerge from the bottom of the steep rocks like walls and quickly mix with the stream. The people call this lowest part of the valley Kocaçay Kısık. The flowing water in the east of Cide is Kısıkiçi Stream. If we look at the structure of the land, it is understood that.
these 'kısık' are cluses. So here we find cluses instead of canyons in typical karst lands.
The main karst events around Cide are seen on the limestone step, which starts from Kayaboğazı in the south of Hondıras and continues to the east of Bartın in an angular fashion. 250 meters of the step, which shows a height of 750 meters compared to the village of Derebucak on its skirt, is in the form of an enormous vertical cornice. Karst pits begin at the upper edge of the cornice. This is also a dense forest of high beeches, hornbeam, oak, ash, elm, and maple trees. The ground is covered by forest vegetation, such as nettles and ferns. Dolines of various depths, large and small, perforated this sloping plateau like a filter, and the limestone rocks were sliced in lapilae, which were covered with stone moss, carved like a pipe in some places, and sharp like a knife—some of the trees of the forest spring out of the dolines like a well. The foot sprains are slipping over overturned, wet tree trunks, mushrooms and moss, getting stuck between sharp lapia. The branch you cling to not to fall is either a rigid and elastic boxwood or a thorny blackberry; one slap, and the other hugs your arm and leg. You slip and stumble at every step, and in the end, you can get away with this karst forest with a bruise on your face, your top is torn, and your shoes pretty battered.

The first cave research in the region also belongs to the author of the above lines, Prof. Cemal Arif Alagöz. This influential geographer writes about Ilgarini in his work published in 1944 with the following lines:

On the way to the Sorkun highland, Ilgarini or Büyükin, whose presence I learned in Cide, is hidden in a dense forest. I can only give a general idea about this cave, which I looked at in the light of the kindling we burned due to the lack of time.
Entering from the 8–10 meters wide and 4–5 meters high methal of Ilgarini, whose mouth is on the south side, and proceeding for 20–25 meters, a wide slit descended to the ground is seen on the right side of the arch-shaped ceiling. (

Fig. 2.20)This entrance, which has cavities on its walls, illuminates a part of the cave. Here, the cave is divided into two: if you go straight ahead, you will find a neat little pool with water in it; Further, in an enlarged part of the cave, a cluster of rough stalactites is seen on the ceiling.
There are pillars and layers of travertine panels formed by coarse stalactite and stalagmites. Starting from this hall, called the 'Chandelier Hall', the floors are always slippery travertine in steps. On the right, enter through a narrow door and return after reaching the cave's end. So far, Ilgarini must be at most 100–150 meters in length. On the way back, the second part of the cave is entered by bending down from a low place opposite the hole in the ceiling. You can go down 15–20 meters from the regular stairs, which they say that there are 24 steps. This is a large, high-ceilinged hall. In the middle of it, there is a tiny building, which is said to be a temple, with the size of 3 x 3.5 meters, built with very carefully hewn stones. The cave changes its direction in this part and descends more or less parallel to the first part above. However, after some progress through the slippery travertines, a deep abyss with a puddle is encountered. To recognize this part of the cave, rope and strong lamps are needed. The underground stream, which initially came out of the large methal, must have descended from the second part of the cave with the narrow surface, under the reasons explaining Yarımburgaz. From the remnants of the pool and the tiny structure, it can be guessed that this place served as a shelter 20–25 years ago. It is necessary to come here again through «exploration» to make an uncomplicated plan of the cave and conduct hydrographic research. There is a rumor among the people that the water lost from here and collected in the Sorkun plateau enters a sinkhole and pours into Kocaçay across the Gömeren village of the Hamitli village of Cide.

Over the years that followed, the cavers of Boğaziçi University Cave Research Club (BÜMAK), having read and been inspired by the aforementioned book, set out to explore the Küre Mountains National Park in 1981. Their primary objective was to locate, explore, and survey the elusive Ilgarini Cave, as mentioned by Alagöz. However, in the surrounding villages, the name "Ilgarini" was unfamiliar to the locals. Eventually, with the aid of a guide from Derebucak

Fig. 2.20 Entrance of Ilgarini Cave (Alagöz, 1944)

Fig. 2.21 Topmeydanı Sinkhole, 1985 (photo M. Albukrek)

Village, they embarked on a challenging 7-h trek through dense jungle to finally reach the cave. After successfully surveying and mapping Ilgarini Cave, caving activities in the region gained momentum.

Within the initial 2 years, the BÜMAK cavers conducted explorations and surveys of Suini, Medil 1, Medil 2, Cıngır, and Sorkun caves, in addition to Ilgarini Cave. In 1982, Ilgarini Cave reached the status of the third deepest cave in Turkey, with a depth of −250 m. The exploration efforts continued, and in 1985, BÜMAK explored the Topmeydanı, Kapaklı, and Döngelyanı sinkholes (Fig. 2.21).

Between 1981 and 1993, BÜMAK cavers discovered a total of 28 caves in the region, successfully surveying each one (Albukrek, 1986, 1993a, 1993b; Ülkümen, 1983).

In 1990, the British LUSS caving team arrived at the Küre Mountains National Park, following a recommendation from Nuri Güldalı, an MTA officer whom they had met during their research in the Taurus Mountains a year earlier. During their 20-day search, the LUSS team explored and mapped several caves in the area. Unaware of BÜMAK's earlier exploration of Ilgarini Cave, the LUSS team re-surveyed and mapped it. Additionally, they discovered and explored the Skull Cave, Snail Cave, Hamitli Cave, Önüne Cave, and Botanical Cave (Holland, 1991). Unfortunately, their report lacks the local names and detailed maps of some of the caves, making the precise locations uncertain.

In 1992, British cavers from the FACC also visited the region, but their expedition was not as fruitful. Despite spending most of their 2-week research efforts on smaller caves like Inönü, they stumbled upon Atak Cave just 4 days before their trip concluded. Although they couldn't fully explore the cave within the limited time, their study proved highly successful (Harris, 1992).

In the same year, the French ADEKS group, which came to the region as part of a general Turkey survey, discovered and surveyed 11 caves around Pınarbaşı and Cide. Even more negatively from the LUSS and FACC teams, since they only gave a code (TU92) to the caves instead of their local names, and there is no detailed information about their locations, it is not possible to understand which caves they are.

One of the descriptions in their report illustrates this issue: *"As for the cave TU 92-51, it would be utopian to try to find it by reading this report! Let us point out that it is located about 25 km east-south-east of the coastal town of Cide, still in the Kastamonu region"* (Wolozan, 1992).

In the vicinity of the villages of Kayabaşı and Karafasıl, where ADEKS conducted their research, no other cave explorations had been done before or after 1992. Nevertheless, we compared the cave maps drawn by this team with all other cave maps to look for similarities and we could not find any. Moreover, the deepest cave they found and explored in the region during this whole study was the aforementioned TU 92–51 with a depth of −45 m.

By 2008, in addition to the caves discovered by BUMAK, a total of 65 caves had been explored in the region by MTA and MAD and with the contributions of LUSS, FACC, and ADEKS teams. Then, 49 additional caves were found throughout our project. In more recent times, 22 more caves have been discovered and explored. As a result, during research done in the area from the beginning to the present, 136 caves were examined and surveyed. As we mentioned above, since they were given different names and their coordinates were not available, only when we examined all the reports and plans did we find that 11 of these caves were duplicates. 125 caves remain after these duplicates are eliminated, and Chap. 3 contains a list of them.

References

Akkemik, Ü., Aytuğ, B., & Güzel, S. (2004). Archaeobotanical and Dendroarchaeological studies in Ilgarini Cave (Pınarbaşı, Kastamonu, Turkey). *Turkish Journal of Agriculture and Forestry, 28*(1), 9–17.

Alagöz, C. (1944). *Türkiye Karst Olayları Hakkında Bir Araştırma (Une étude sur les phénomènes karstiques en Turquie)*. Ankara.

Albukrek, M. (1986). Kastamonu Bölgesi Araştırmaları II. *DELTA, 2*, 5–15.

Albukrek, M. (1993a). Kastamonu Bölgesi Araştırmaları III. *DELTA, 3*, 29–43.

Albukrek, M. (1993b). Kastamonu Bölgesi Araştırmaları IV. *DELTA, 4*, 15–25.

Ayengin, N. (2020). Kahintepe: Karadeniz'de Bir Tapınım Alanı. *Aktüel Arkeoloji, 76*(12–13).

Aylar, F., Zeybek, H., & Dinçer, H. (2018). Karstification and Karstic shapes between the rivers Devrekâni and Aydos of the Küre Mountains. *Eastern Geographical Review, 23*(39), 1–24.

Burney, C. (1956). *Northern anatolia before classical times* (pp. 179–203). Anatolian Studies.

Derman, A. (1990). *Batı Karadeniz Bölgesi'nin Geç Jura-Erken Kretase'deki Jeolojik Evrimi.* Paper presented at the Türkiye 8. Petrol Kongresi, Ankara.

Derman, A. (1995). *Cretaceous Megabreccias and Other Mass Flow Deposits in the Eastern Ulus Basin. Western Black Sea Region of Türkiye (unpublished).* PhD, Middle East Technical University, Ankara.

Dokay. (2018). *Bartın Mağaraları Hakkında Araştırma Raporu.* BAKKA Batı Karadeniz Kalkınma Ajansı.

Düring, B., & Glatz, C. (2015). *Kinetic landscapes the cide archaeological project: Surveying the Turkish Western Black Sea region.* De Gruyter Open Ltd.

Düring, B., Glatz, C., & Şerifoğlu, T. (2012). The cide archaeological project 2011: Third preliminary report. *Anatolia Antiqua, 20*, 167–175.

Emre, K., & Çınaroğlu, A. (1993). A group of metal vessels from Kınık-Kastamonu. In M. Mellink, E. Porada, & T. Özgüç (Eds.), *Aspects of art and iconography: Anatolia and its neighbors, studies in honor of Nimet Özgüç* (pp. 675–713). Ankara.

Gand, G., Tüysüz, O., Steyer, J. S., Allain, R., Sakınç, M., Sanchez, S., Şengör, A. M. C., & Şen, Ş. (2011). New Permian tetrapod footprints and macroflora from Turkey (Çakraz formation, northwestern Anatolia): Biostratigraphic and palaeoenvironmental implications. *Comptes Rendus Palevol, 10*, 617–625.

Gedik, İ., & Aksay, A. (Cartographer). (2002). Geological Map of Zonguldak E29.

Genç, E. (2005). *Yeni buluntular ışığında Kastamonu-Kınık kazıları sonuçları.* PhD, Akdeniz Üniversitesi Sosyal Bilimler Enstitüsü, Antalya.

Glatz, C., Düring, B., Şerifoğlu, S., & Emre, T. (2011). The cide archaeological project 2010: Second preliminary report. *Anatolia Antiqua, 19*, 279–288.

Glatz, C., & Matthews, R. (2005). Anthropology of a frontier zone: Hittite-Kaska relations in late bronze age North-Central Anatolia. *Bulletin of the American Schools of Oriental Research, 339*, 47–65.

Güney, K. (2021). *Küre Dağları Milli Parkı'nın (Kastamonu) Riparian Vejetasyonunun Sintaksonomik Analizi.* (PhD), Ankara Üniversitesi, Ankara.

Güney, K., Geven, F., Bingöl, M., Çeter, T., Altuner, E., Aydın, M., & Sıvacıoğlu, A. (2012). *Küre Dağları Milli Parkı Biyoçeşitliliği: Ekosistem ve Flora.* Paper presented at the Biyolojik Çeşitlilik Sempozyumu Ankara.

Harris, S. (1992). FACC in Turkish delight. *Caves and Caving, 62.*

Holland, M. (1991). LUSS Turkey 90 expedition report. *Caves and Caving, 52*, 24–27.

Kazancı, N., Suludere, Y., Mülazımoğlu, N., Tuzcu, S., Mengi, H., & Hakyemez, H. (2009). *Küre Dağları Milli Parkı ve Çevresi Jeositleri.* Ankara.

Kökten, K. (1948). 1947 Yılı Tarih Öncesi Araştırmaları. *Belleten, 12*(45), 223–226.

Kuzucuoğlu, C., Marro, C., Özdoğan, A., & Tibet, A. (1997). Prospection archéologique franco-turque dans la région de Kastamonu (Mer Noire). Deuxième rapport préliminaire. *Anatolia Antiqua, 5*, 275–306.

Marro, C., Özdoğan, A., & Tibet, A. (1996). Prospection archéologique franco-turque dans la région de Kastamonu (Mer Noire). Premier rapport préliminaire. *Anatolia Antiqua, 4*, 273–290.

Marro, C., Özdoğan, A., & Tibet, A. (1998). Prospection archéologique franco-turque dans la région de Kastamonu (Mer Noire). Troisième rapport préliminaire. *Anatolia Antiqua, 6*, 317–335.

Okay, A. (2008). Geology of Turkey: A synopsis. *Der Anschnitt, 21*, 19–42.
Özdoğan, A., Marro, C., & Tibet, A. (1996). *Kastamonu Yüzey Araştırmaları (1995)*. Paper presented at the 14. Araştırma Sonuçları Toplantısı, Ankara.
Özdoğan, A., Marro, C., & Tibet, A. (1997). *Kastamonu Yüzey Araştırması 1996 Yılı Çalışmaları*. Paper presented at the 15. Araştırma Sonuçları Toplantısı, Ankara.
Özdoğan, A., Marro, C., & Tibet, A. (1998). *Kastamonu Yüzey Araştırmaları 1997 Yılı Çalışmaları*. Paper presented at the 16. Araştırma Sonuçları Toplantısı, Ankara.
Şengör, A. (2000). Jeolojik Takvim. *Cogito, 22*.
Şengör, A., Tüysüz, O., Imren, C., Sakınç, M., Eyidoğan, H., Görür, N., et al. (2005). The North Anatolian fault: A new look. *Annual Review of Earth and Planetary Sciences, 33*, 37–112.
Şenyürek, M. (1944). Anadolu'da Bulunan îki Paleolitik Alete Dair bir Not. *Dil Tarih ve Coğrafya Fakültesi Dergisi, 2*(2), 349–351.
Sirel, E. (1991). Cideina, a new foraminiferal genus from the Maastrichtian limestone of the Cide region (North Turkey). *Mineral Research and Exploration Bulletin (MTA), 112*, 65–70.
Summerer, L. (2017). *Pompeiopolis-Taşköprü. 2000 years from Metropolis to county town*. Ege Yayıncılık.
Summerer, L., & von Kienlin, A. (2013). Pompeiopolis. The Metropolis of Paphlagonia. In *L'Anatolie des peuples, des cités et des cultures (IIe millénaire av. J.-C. - Ve siècle ap. J.-C.)* (Vol. 2, pp. 115–126). Institut des Sciences et Techniques de l'Antiquité.
Tekebaş, S. (2017). *Küre Dağları Milli Parkı'nın Bartın Bölümünde Bulunan Zoni Yaylası ve Etrafının Florası*. BA, Bartın Üniversitesi, Bartın.
Tunçkol, B., Aksoy, N., & Yaşayacak, H. (2020). Classification of endemic plants and priority conservation areas in Küre Mountains National Park (Kastamonu section). *Journal of Forestry, 16*(2), 148–160.
Tunçkol, B., Aksoy, N., & Yaşayacak, H. (2022). Küre Dağları Milli Parkı'nın Florası (Kastamonu Bölümü). *Eurasian Journal of Forest Science, 10*(3), 72–107.
Tunoğlu, C., & Bardet, N. (2006). Mosasaurus hoffmanni Mantell, 1829: First marine reptile discovery from late cretaceous of Türkiye. *Türkiye Jeoloji Bülteni, 49*(1).
Tüysüz, O. (2022). Geology of the Kurucaşile - Cide region, NW Türkiye. *Bulletin of the Mineral Research and Exploration, 167*, 149–178.
Ülkümen, O. (1983). Kastamonu Bölgesi Araştırmaları I. *DELTA, 1*, 1–9.
Ünal, A. (1999). A hittite mycenaean type B sword from the vicinity of Kastamonu, Northwest Turkey. *Bulletin of the Middle Eastern Culture Center in Japan, XI*, 207–226.
Wolozan, D. (1992). *Recherches Spéléologiques en Turquie (4ème et 5ème campagnes de l'ADEKS)* (Vol. 3).
Yıldırım, Ş. (2013). *The Rescue Excavation of the Selmanlı Tumulus in Kastamonu*. Paper presented at the The Danubian Lands between the Black, Aegean and Adriatic Seas (7th Century BC—10th Century AD). Proceedings of the Fifth International Congress on Black Sea Antiquities, Belgrade.
Yıldırım, T., Bal, N., & Dardeniz, G. (2021). Symbolism in 13th-century bc Hittite metallurgy: The Kastamonu-Kınık (Turkey) metal hoard, again. *Journal of Near Eastern Studies, 80*(1), 31–57.

3 Caves of Küre Mountains National Park

3.1 Introduction

We were 20 km north of Kastamonu province's Pınarbaşı district, near Kazla Village, on the edge of the Küre Mountains National Park, and there cannot possibly be any undiscovered cave in this area. For years, cavers have been exploring this region, looking at every basin from Kazla to Dizdarlı villages. Five caves were discovered even in this small region. However, at the same spot, a brand-new cave appeared in front of us—a previously unknown sinkhole. The unexpected discovery of a new cave near Kazla Village underscores the continuous potential for uncharted cave systems, challenging prior assumptions about the comprehensiveness of exploration efforts in the region.

When we began working on the cave inventory project of Küre Mountains National Park, we intended to divide it into five sections and proceed from one section to the next after finishing the first. As the first section, we chose the part of the National Park located northwest of the Kastamonu-Pınarbaşı district. We chose this area because it was relatively easy to reach and had been explored many times before. There couldn't be many undiscovered caves in this area after 25 years of exploration started by Boğaziçi University Cave Research Club (BÜMAK) in the early 1980s. We would focus on the unexplored areas of Küre Mountains National Park after exploring the last few small undiscovered caves here.

It was understood from the first trip how wrong we were. Around Pınarbaşı, from Topmeydanı to Ilıca, from Karafasıllı to Demirci, we encountered new caves everywhere. As if that were not enough, new branches were found in formerly explored caves. Despite the impressive number of caves already documented, the discovery of new branches in previously explored caves and the identification of previously unknown sinkholes highlight the dynamic and ever-evolving nature of cave systems in the Küre Mountains National Park.

As explained in Chap. 2, cave explorations in this region started with Alagöz in 1944 (Alagöz, 1944). After reading and getting inspired by his book, the cavers of BÜMAK embarked on an exploration of the Küre Mountains' caves in the subsequent years. Their primary objective was to locate, explore, and survey the elusive Ilgarini Cave, as mentioned by Alagöz. After successfully surveying and mapping Ilgarini Cave, caving activities in the region gained momentum. BÜMAK cavers explored and surveyed the caves of Suini, Medil 1, Medil 2, Cıngır, and Sorkun during the first 2 years of their expedition. They also explored the sinkholes at Topmeydanı, Kapaklı, and Döngelyanı in 1985. In all, they found and successfully surveyed 28 caves in the area between 1981 and 1993 (Albukrek, 1986, 1993a, 1993b; Ülkümen, 1983). The exploration and survey of these 28 caves by BÜMAK laid a solid foundation for subsequent research, establishing a baseline for understanding the subterranean landscape in the Küre Mountains National Park.

The British LUSS caving team explored and mapped six caves at the Küre Mountains National Park in 1990. They conducted a new survey and mapping of Ilgarini Cave without knowing about BÜMAK's previous exploration. They also found and surveyed Snail Cave, Hamitli Cave, Önüne Cave, Botanical Cave, and Skull Cave (Holland, 1991).

Regrettably, their report is devoid of precise locations due to the absence of detailed maps and local names for some of the caves. The absence of precise locations in the reports of the LUSS caving team poses a challenge to future researchers, emphasizing the importance of detailed mapping and documentation for the benefit of the scientific community.

The area was also visited by British cavers from the FACC in 1992, although their expedition was not as successful. Just 4 days before their 2-week expedition came to an end, they discovered Atak Cave, despite having focused the majority of their research efforts on smaller caves like Inönü. The challenges faced by this team highlight the unpredictable nature of cave discoveries, with Atak Cave being explored just days before the conclusion of their expedition (Harris, 1992). During a general survey of Turkey in the same year, the French ADEKS group visited the area and found and

A. Yamaç, E. Tok, *Caves and Karst of Küre Mountains National Park, Turkey*, Cave and Karst Systems of the World,
https://doi.org/10.1007/978-3-031-80142-6_3

surveyed 11 caves near Pınarbaşı and Cide. Even worse than the LUSS and FACC teams, they only gave a code to the caves instead of using their local names, and there is no detailed information about their locations.

In addition to BÜMAK's comprehensive research, 65 caves were found and explored in the region as of 2008, with the contributions of MTA, MAD, LUSS, FACC, and ADEKS teams.

Some of these caves were duplicates, and others required rediscovery due to missing exact coordinates or unambiguous directions. Unfortunately, some of the caves could not be found.

In future exploration attempts, the persistent difficulty in locating certain caves because of missing coordinates or unclear directions emphasizes the necessity of thorough record-keeping and the development of comprehensive mapping techniques (Albukrek, 1986, 1993a, 1993b; Harris, 1992; Holland, 1991; Ülkümen, 1983; Wolozan, 1992).

After these studies mentioned above, during our project, we visited the Küre Mountains National Park 14 times in 2 years. The collaborative approach taken during the project's 14 visits to the national park over 2 years facilitated the discovery of an additional 49 caves, underscoring the importance of continuous and dedicated exploration efforts. The diverse range of caves explored, from sinkholes to deep caverns, showcases the dynamic geological processes at play in the Küre Mountains National Park, providing valuable insights into the region's subterranean landscape.

Twenty-two more caves were discovered and surveyed in the area during the more recent studies conducted by multiple teams. The "Cide Archaeological Project" (CAP) is one such study. Seven caves were archaeologically surveyed as part of this project, which was the first archaeological survey of the area. The interdisciplinary approach taken by this project not only expanded the known inventory of caves but also provided archaeological insights, further enriching the scientific understanding of the region (Düring et al., 2012; Düring & Glatz, 2015; Glatz et al., 2011). The discoveries made in these caves are covered in more detail below.

As a result of all these research and studies carried out over the years, a total of 136 caves were explored and surveyed in and around Küre Mountains National Park. The discovery of these caves in the region through years of research underscores the vast potential for further exploration in the expansive limestone area. The accumulated knowledge from various expeditions led to the identification and correction of duplicate cave entries, resulting in a refined database of 125 unique caves (Figs. 3.1 and 3.2).

The geographical distribution of these 125 caves is interesting. While most of the caves are located around the plateau where the national park is located, some are even outside the buffer zone, and only a very few caves have been discovered on the plateau. The most important reason for this is that access to the plateau can only be made from two points, and the top of the plateau is a real jungle. The limited accessibility to the plateau, coupled with its dense vegetation, has contributed to the discovery of only a small fraction of caves in the core part of the national park, highlighting the challenges faced by researchers in certain topographical areas. Only 12 caves have been discovered in this 370 km^2 core part of the national park. Five of them are in the Armutlu Plateau, the only point that can be reached by car from the north of the plateau (Fig. 3.3).

Discovered in an area of more than 1170 km^2, which is almost entirely limestone, these caves must be a tiny fraction of the existing number. There must be hundreds of other caves waiting to be discovered in such a large limestone area. On the other hand, we already know very well that this is the case. Because when this project ended on the set date, we had 92 more caves about which we had information but could not investigate. The existence of these 92 caves with available information that have not yet been explored emphasizes the need for sustained efforts in cave exploration to uncover the full extent of the subterranean landscape. The vast limestone expanse in the western and central Black Sea region presents an intriguing prospect for the existence of numerous undiscovered caves.

There are limestone layers of different ages in large areas in the western and central Black Sea region. Although these horizontal limestone layers allowed the formation of long caves such as Kızılelma and Gökgöl in the Western Black Sea region, examples of similar length have not yet been found in the central Black Sea region. This can be explained by the inadequacy of research.

On the other hand, due to the geomorphological formation of this region, the thickness of the limestone layers is too low to be compared with the Mediterranean region. The scarcity of deep caves in Küre Mountains National Park correlates with the thickness of limestone layers, emphasizing the role of geological factors in cave formation. The listing of the longest and deepest caves explored in the region provides a quantitative perspective on the notable dimensions achieved by certain cave systems, showcasing the diversity in cave morphology.

The longest caves explored in the region:

1. Topmeydanı Sinkhole 1470 m
2. Atak Cave 1365 m
3. Eşek Çukuru Sinkhole 1231 m
4. Ilgarini Cave 858 m
5. Gurbettepe Sinkhole 710 m
6. Kurtyolu 1 Sinkhole 586 m

The deepest caves explored in the region:

1. Dağlı Sinkhole −279 m
2. Ilgarini Cave −250 m

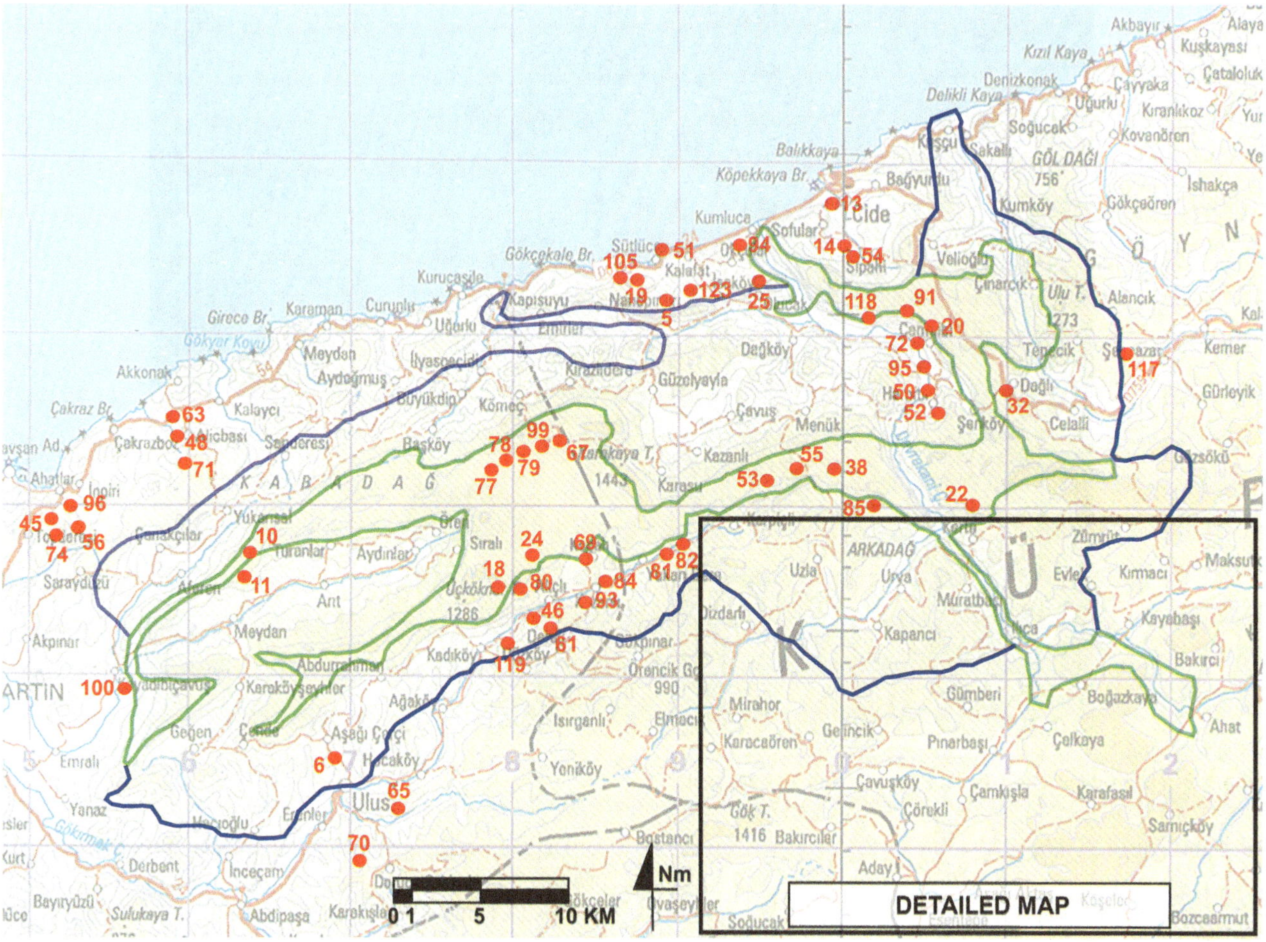

Fig. 3.1 Location map showing all the known caves in and around Küre Mountains National Park. The green line is the border of the national park, and the blue line is the buffer zone (base map HGM-elaboration A. Yamaç)

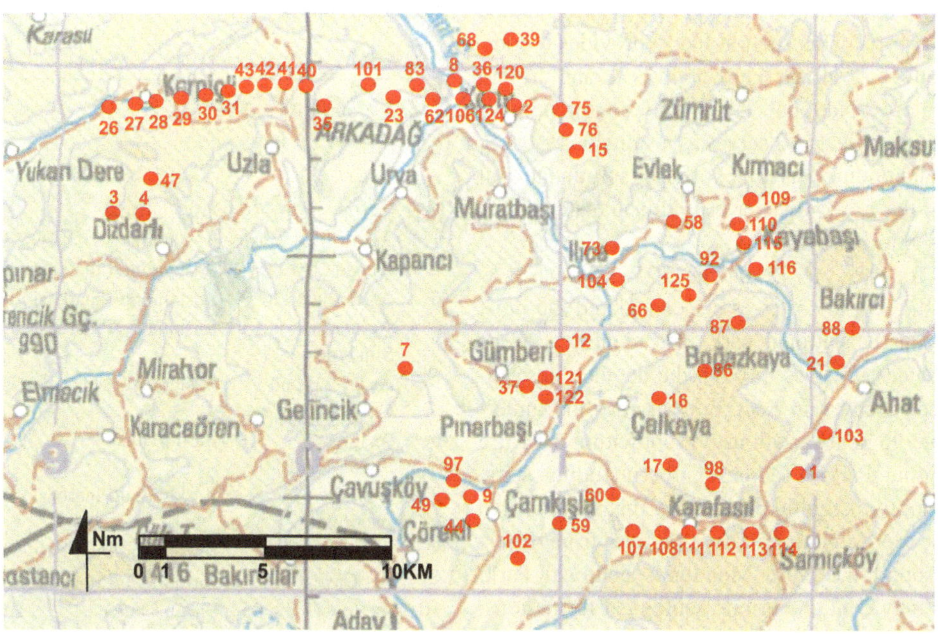

Fig. 3.2 Detailed section of Fig. 3.1 (base map HGM-elaboration A. Yamaç)

Fig. 3.3 The road towards Armutlu Plateau (photo M. Eğrikavuk)

3. Gurbettepe Sinkhole −243 m
4. Kapaklı Sinkhole −204 m
5. Sorkun Sinkhole −200 m
6. Çengelkuyu Sinkhole −199 m

Section 3.2 contains a list of all 125 caves found in the region, including a few that are merely identified by name and location in reports. This section likely provides a comprehensive inventory of all the known caves in and around the Küre Mountains National Park. The comprehensive inventory presented in Sect. 3.2 serves as a valuable resource for researchers and enthusiasts alike, offering a foundational understanding of the 125 known caves within and around the Küre Mountains National Park.

Section 3.3 focuses on 45 of the 125 known caves described in detail. These 45 caves are considered the most important caves in the national park. The identification of structural differences in caves through the detailed descriptions of 45 selected caves in Sect. 3.3 enriches the understanding of the geological variations present within the Küre Mountains National Park. The reason for including some smaller or seemingly insignificant caves in this section is to provide a diverse representation of cave samples from different locations around the Küre Mountains National Park. These locations may contain various geological formations, and including these caves allows for a presentation of the structural differences in terms of caves within the park.

The tag numbers associated with cave names in Sect. 3.3 provide a navigational guide, correlating with detailed maps in Figs. 3.1 and 3.2 for precise location identification. The scientific criteria applied in translating the legends of the cave plans into English ensure a standardized presentation that facilitates international cooperation and understanding. On the other hand, there have been no other changes made to these cave plans.

3.2 List of Caves

No	Cave	Prov./District	Village/Region
1	Ağzı Büyük Cave	Kastamonu/Azdavay	Sarnıç
2	Alman Cave	Kastamonu/Pınarbaşı	Sümenler
3	Arpatarlası Cave	Kastamonu/Pınarbaşı	Dizdarlı/Zelle
4	Arpatarlası Sinkhole	Kastamonu/Pınarbaşı	Dizdarlı/Zelle
5	Aşağı Baliönü Cave	Kastamonu/Cide	Gideros
6	Aşağıçerçi Sinkhole	Bartın/Ulus	Aşağıçerçi
7	Atak Cave (Uçurum)	Kastamonu/Pınarbaşı	Çavuş
8	Ayakkabı Cave	Kastamonu/Pınarbaşı	Sümenler
9	Ayı Cave	Kastamonu/Pınarbaşı	Güvercinlik
10	Ayıgölü Sinkhole	Bartın Center	Arıt/Çöpbey/Zoni
11	Bacaksızın Cave	Bartın Center	Arıt/Çöpbey
12	Balcak Cave	Kastamonu/Pınarbaşı	Ilıca/Esen Mah
13	Ballıcı Cave	Kastamonu/Cide	Abdülkadir

(continued)

No	Cave	Prov./District	Village/Region
14	Botanik Cave	Kastamonu/Cide	Sırakaya/Loç
15	Buz Sinkhole	Kastamonu/Pınarbaşı	Kerte
16	Buzluk Grotto	Kastamonu/Pınarbaşı	Çuhadar
17	Buzluk Cave	Kastamonu/Pınarbaşı	Karafasıl/Sarma
18	Büyük Kaya Cave	Bartın/Ulus	Alıçlı/Büyükkaya
19	Çadır Cave (Fındık)	Kastamonu/Cide	Gideros
20	Çamdibi Cave	Kastamonu/Cide	Çamdibi
21	Çatak Ağzı 1,2,3	Kastamonu/Azdavay	Ahatköy
22	Çengelkuyu Sinkhole	Kastamonu/Pınarbaşı	Kerte
23	Çıkrıkkapı Sinkhole	Kastamonu/Pınarbaşı	Sümenler
24	Çıngıraklı Sinkhole	Bartın/Ulus	Alıçlı
25	Çıngırık Sinkhole	Kastamonu/Cide	Okçular
26	Çovurma Tepe Sinkhole	Kastamonu/Pınarbaşı	Uzla
27	Çovurma Tepe Cave 1	Kastamonu/Pınarbaşı	Uzla
28	Çovurma Tepe Cave 2	Kastamonu/Pınarbaşı	Uzla
29	Çovurma Tepe Cave 3	Kastamonu/Pınarbaşı	Uzla
30	Çovurma Tepe Cave 4	Kastamonu/Pınarbaşı	Uzla
31	Çovurma Tepe Cave 5	Kastamonu/Pınarbaşı	Uzla
32	Dağlı Sinkhole	Kastamonu/Şenpazar	Dağlı
33	Derebağ Village Cave	Kastamonu/Cide	Derebağ
34	Dikenlikaya	Bartın Center	İnkumu
35	Dikilitaş Cave (Skull)	Kastamonu/Pınarbaşı	Yamanlar
36	Döngelyanı Sinkhole	Kastamonu/Azdavay	Kazla
37	Düşenkaya Cave	Kastamonu/Pınarbaşı	Demirci
38	Ejder Cave	Kastamonu/Pınarbaşı	Yamanlar
39	Ejder Cave 2	Kastamonu/Pınarbaşı	Valla Canyon
40	Eşek Çukuru Cave 1	Kastamonu/Pınarbaşı	Uzla
41	Eşek Çukuru Cave 2	Kastamonu/Pınarbaşı	Uzla
42	Eşek Çukuru Sinkhole	Kastamonu/Pınarbaşı	Uzla
43	Gavur Cave	Kastamonu/Pınarbaşı	Uzla
44	Gökdeli Cave (Mızrak)	Kastamonu/Pınarbaşı	Güvercinlik
45	Gökpınar Cave	Bartın/Amasra	Makaracı
46	Gölcük Cave	Bartın/Ulus	Alıçlı
47	Gurbettepe Sinkhole	Kastamonu/Pınarbaşı	Dizdarlı Zelle
48	Gürcüoluk Cave	Bartın/Amasra	Makaracı
49	Gürlik Cave	Kastamonu/Pınarbaşı	Güvercinlik
50	Hacet Kayası Cave	Kastamonu/Cide	Karakadı
51	Hamam Cave	Kastamonu/Cide	Gideros
52	Hamitli Cave	Kastamonu/Cide	Hamitli
53	İğnedeligi Cave (Salyangoz)	Kastamonu/Pınarbaşı	Yamanlar

No	Cave	Prov./District	Village/Region
54	Ilahi Cave	Kastamonu/Cide	Sırakaya/Loç
55	Ilgarini Cave	Kastamonu/Pınarbaşı	Yamanlar
56	İnderesi Cave	Bartın/Amasra	Makaracı
57	İnkumu Cave	Bartın Center	İnkumu
58	İnönü Grotto	Kastamonu/Pınarbaşı	Ilıca/Göktepe
59	İnönü Cave	Kastamonu/Pınarbaşı	Güvercinlik
60	İntürbesi Cave	Kastamonu/Pınarbaşı	Çalkaya
61	Kadıharmanı Sinkhole	Bartın/Ulus	Alıçlı
62	Kapaklı Sinkhole	Kastamonu/Pınarbaşı	Sümenler/Kazla
63	Karakaçak Cave	Bartın/Amasra	Karakaçak
64	Karakış Cave	Kastamonu/Pınarbaşı	
65	Karain Cave	Bartın/Ulus	Eldeş
66	Kaval Sinkhole	Kastamonu/Pınarbaşı	Boğazkaya
67	Kayaarkası Sinkhole—Cave	Kastamonu/Cide	Armutlu
68	Kayadibi Sinkhole	Kastamonu/Pınarbaşı	Sümenler/Valla
69	Kemerli Cave	Bartın/Ulus	Kemerli
70	Kemikli Cave	Bartın/Ulus	Doruca
71	Kirenlik Cave	Bartın/Amasra	İncüvez
72	Kılıçlı Cave	Kastamonu/Cide	Çamdibi
73	Kınakayası Cave	Kastamonu/Pınarbaşı	Ilıca/Esen
74	Kocareis Cave	Bartın/Amasra	Makaracı
75	Kokurdan 1 Sinkhole	Kastamonu/Pınarbaşı	Kerte
76	Kokurdan 2 Sinkhole	Kastamonu/Pınarbaşı	Kerte
77	Kurtyolu Sinkhole	Kastamonu/Cide	Armutlu
78	Kurtyolu 1 Sinkhole	Kastamonu/Cide	Armutlu
79	Kurtyolu 2 Cave	Kastamonu/Cide	Armutlu
80	Kuşaklı Cave	Bartın/Ulus	Alıçlı/Büyükkaya
81	Kuşkayası 1 Cave	Bartın/Ulus	Yukarıdere/Göksu
82	Kuşkayası 2 Cave	Bartın/Ulus	Yukarıdere/Göksu
83	Küçük Çıkrıkkapı Sinkhole	Kastamonu/Pınarbaşı	Sümenler
84	Manastır Sinkhole	Bartın/Ulus	Köklü
85	Mantar Cave	Kastamonu/Pınarbaşı	Sümenler
86	Medil 1 Cave	Kastamonu/Azdavay	Kurtgirmez/Mus
87	Medil 2 Cave	Kastamonu/Azdavay	Kurtgirmez/Örencik
88	Mendiltaş Cave	Kastamonu/Azdavay	Karakuşlu
89	Mugada 1 Cave	Bartın Center	Mugada
90	Mugada 4 Cave	Bartın Center	İnkumu
91	Muna Türbe Cave	Kastamonu/Cide	Atköprü
92	Mürsili Cave	Kastamonu/Pınarbaşı	Horma Canyon
93	Nalçak Moğarı	Bartın/Ulus	Kozanlı/Yukarıdere
94	Okçular Cave	Kastamonu/Cide	Gideros
95	Önüne Cave	Kastamonu/Cide	Çalköy
96	Perili Cave	Bartın/Amasra	Makaracı
97	Sıçan Delik Cave	Kastamonu/Pınarbaşı	Güvercinlik
98	Sırt Sinkhole	Kastamonu/Pınarbaşı	Karanfil
99	Sinekli Cave	Kastamonu/Cide	Armutlu
100	Sipahiler Cave	Bartın Center	Sipahiler
101	Sorkun Sinkhole	Kastamonu/Pınarbaşı	Yamanlar
102	Subaşı Cave	Kastamonu/Pınarbaşı	Kurtlugelik
103	Suini Cave	Kastamonu/Azdavay	Sarnıç
104	Şelale Cave	Kastamonu/Pınarbaşı	Ilıca

(continued)

No	Cave	Prov./District	Village/Region
105	Şinasi Cave (Gideros 1)	Kastamonu/Cide	Gideros
106	Topmeydanı Sinkhole	Kastamonu/Pınarbaşı	Sümenler
107	TU 922	Kastamonu/Pınarbaşı	Karafasıl
108	TU 92-3	Kastamonu/Pınarbaşı	Karafasıl
109	TU 92-4	Kastamonu/Azdavay	Kayabaşı
110	TU 92-5	Kastamonu/Azdavay	Kayabaşı
111	TU 92-100	Kastamonu/Pınarbaşı	Karafasıl
112	TU 92-101	Kastamonu/Pınarbaşı	Karafasıl
113	TU 92-102	Kastamonu/Pınarbaşı	Karafasıl
114	TU 92-103	Kastamonu/Pınarbaşı	Karafasıl
115	TU 92-104	Kastamonu/Azdavay	Kayabaşı
116	TU 92-106	Kastamonu/Azdavay	Kayabaşı
117	TU 92-51	Kastamonu/Şenpazar	Şenpazar
118	Uçakdibi Cave	Kastamonu/Cide	Atköprü
119	Ulukaya Cave	Bartın/Ulus	Ulukaya
120	Ürkülük Sinkhole	Kastamonu/Pınarbaşı	Sümenler
121	Yarasa Cave	Kastamonu/Pınarbaşı	Demirci
122	Yarasa 2 Cave	Kastamonu/Pınarbaşı	Demirci
123	Yukarı Baliönü Cave	Kastamonu/Cide	Gideros
124	Yukarı Döngelyanı Sinkhole	Kastamonu/Pınarbaşı	Sümenler
125	Yumruçal Cave	Kastamonu/Pınarbaşı	Horma Canyon

3.3 45 Significant Caves

3.3.1 Aşağı Baliönü Cave (No. 5)

Asagi Baliönü Cave is located in the southwest of Gideros Bay of Kurucaşile, Bartın. The cave is located at the bottom of the Upper Cretaceous limestone wall to the west of the Akgöynük Polje, and there are limestone reefs stacked on top of each other in the region.

The sinkhole, which drains all the water of a rather large polje, has a gigantic entrance of 38 m but is only 91 m long. Mud piles are reaching up to 3 m in height in the large chamber just behind the entrance. In the mud, the water has created a new bed for itself. After a short distance, it narrows and turns into an active cave gallery, but then it ends by a siphon. Intense biological diversity was detected in both the muddy and active parts of this cave.

The cave was explored and mapped by MTA in May 2003 in order not to be damaged by a marble quarry to be established in the region (Nazik & Törk, 2003a). It is on the list of registered natural protection sites prepared by the Ministry of Culture and Tourism (Figs. 3.4 and 3.5).

3.3.2 Atak Cave (No. 7)

The cave is situated on the rock wall approximately 100 m from the point where the road terminates, to the north of Pınarbaşı, Çavuş Village. It is one of the longest caves in the region, with a length of 1,365 m. In terms of its structural features and findings, it is one of the most important caves in the region. The cave was developed horizontally within the limestone of the Upper Jurassic-Lower Cretaceous age and in the form of mélanges in this region. The initial 300–400 m of the cave, which developed in the clastic limestones, is filled with depressions, branches, and rock piles. The cave's narrow passages reach the canyon wall, through which the Paşaçayırı Stream flows, and form different entrances to the cave at these points. Conversely, the principal branch of the cave becomes relatively unadorned, with a single gallery from the outset of the descent and continuing in this manner until the conclusion (Fig. 3.6). The elevation, which extends down a short distance from the entrance of the cave and is reached in the last parts of the cave, is almost the same as the stream flowing outside. At these points, muddy ground is encountered with the water leaking from the outside.

The research on the cave, which was first found by the British FACC team that came to the region in 1992, was incomplete (Harris, 1992). The survey was completed in 2008 and published in the first issue of OBRUK (Yamaç, 2008) (Fig. 3.7).

There is a very rich biodiversity in the cave, from 15 different species of insects to several hundred bats and mammals such as *Dryomys nitedula* and *Glis glis* that hibernate here. A wildcat (*Felis sylvestris*) skeleton was also found. Although the entrance parts of the cave are not very rich in terms of formation, there is a section with very beautiful formations in the further parts of the main gallery, and it needs to be taken under immediate protection.

Atak Cave's most intriguing discoveries are human skeletons and pottery fragments found in the gallery beyond a narrow passage at the cave's end (Fig. 3.8). Experts have dated the pottery to the Middle to Early Bronze Age. Additionally, the gallery in this section of the cave is blocked and has no external connection. If these individuals did not arrive at this location by traversing a 1,300-meter cave using the means available 3,000 years ago (and with all the pottery found), or if they did not transport their deceased to this location, it is likely that an entrance from the surface in this area has collapsed and become obstructed over the past several millennia.

3.3.3 Ayıgölü Sinkhole (No. 10)

The Ayıgölü Sinkhole is situated on the Zoni Plateau, in the northern region of Çöpbey Village in Bartın. The depth of the

Fig. 3.4 The entrance of Aşağı Baliönü Cave (photo Ç. Çankırılı)

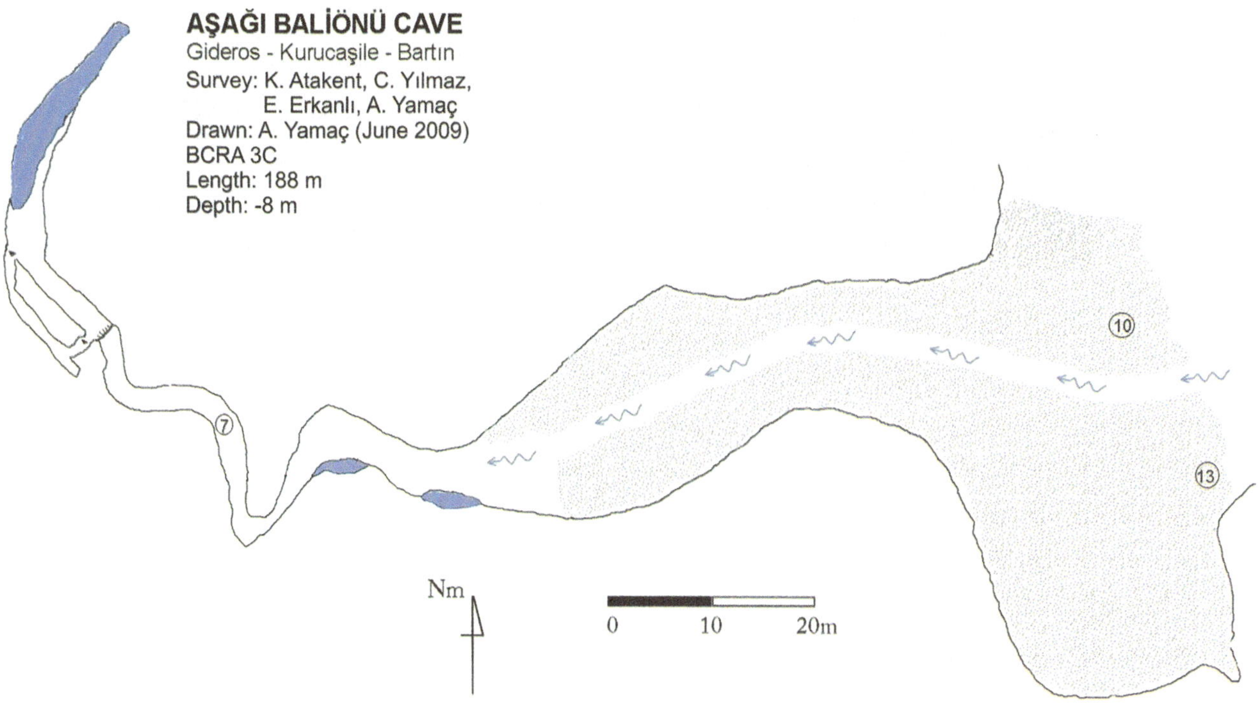

Fig. 3.5 Aşağı Baliönü Cave plan and extended profile (drawn A. Yamaç)

sinkhole is −31 m, with a total length of 233 m (Fig. 3.9). The other entrance to the cave and numerous tributaries are obstructed by alluvium. It is likely that the stream, which has sunk into the sinkhole, has formed a new bed as a result of tectonic movements, with some other branches becoming vadose. Along the main gallery, there are small ponds between the rock piles that have been carried in by water (Fig. 3.10).

It is likely that the karst aquifers in the plain were formed by at least some of the waters that submerged this limestone plateau. However, the thick vegetation makes it challenging to explore these sinkholes. One such example is the Ayıgölü Sinkhole, which can be found in the Zoni Plateau, where there is relatively little vegetation. It is postulated that the water entering this cave, which terminates in a siphon, originates from the nearby Bacaksızın Cave (No. 11). The distance between the two caves is 1.7 km, and the elevation difference is 300 m.

3.3.4 Ballıcı Cave (No. 13)

The Ballıcı Cave is a large cave situated in the Abdülkadir Village of Cide, Kastamonu. The cave has been identified as a site of human occupation during several historical periods. Its documentation was conducted during the 2010 archaeo-

Fig. 3.6 Atak Cave the main gallery (photo D. Lee)

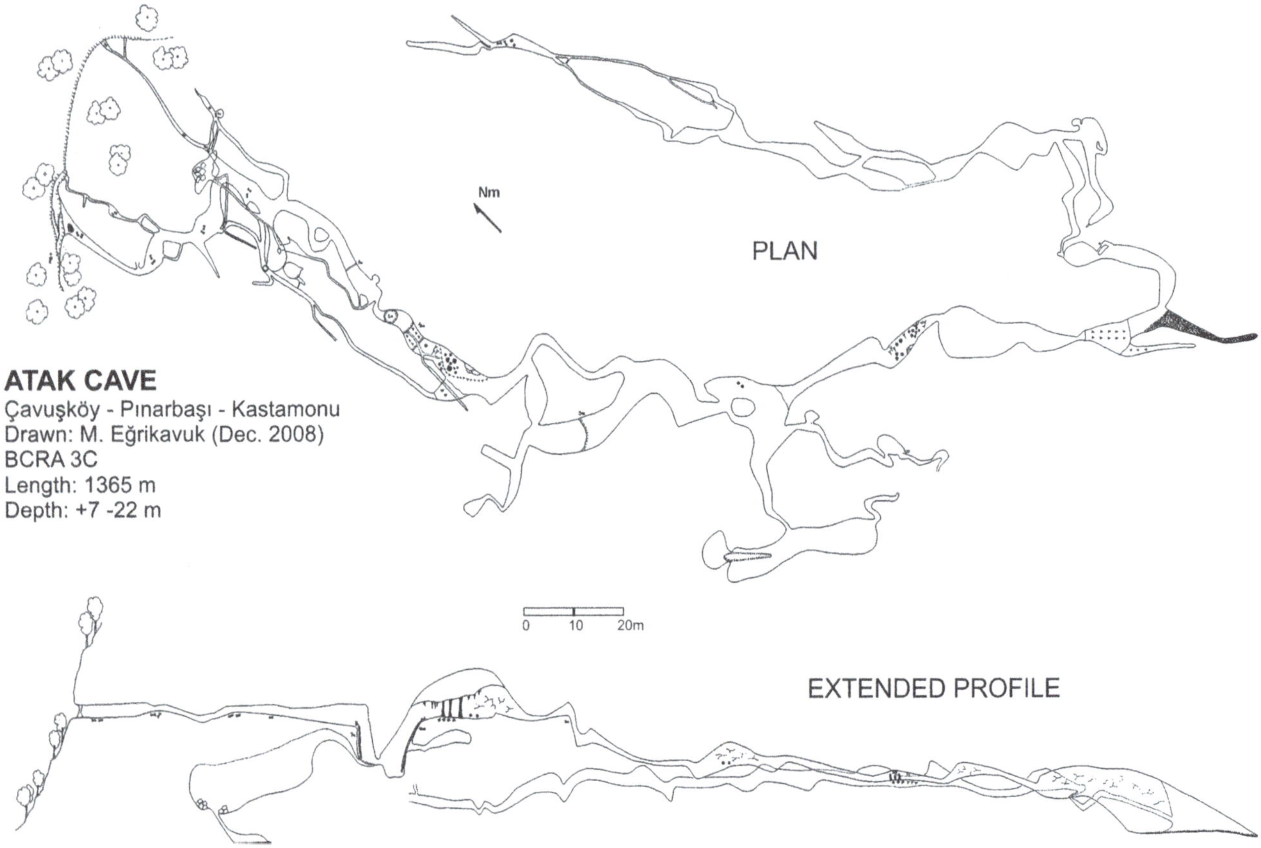

Fig. 3.7 Atak Cave plan and extended profile (drawn M. Eğrikavuk)

Fig. 3.8 Human skeletons in the last gallery (photo S. Bahadır)

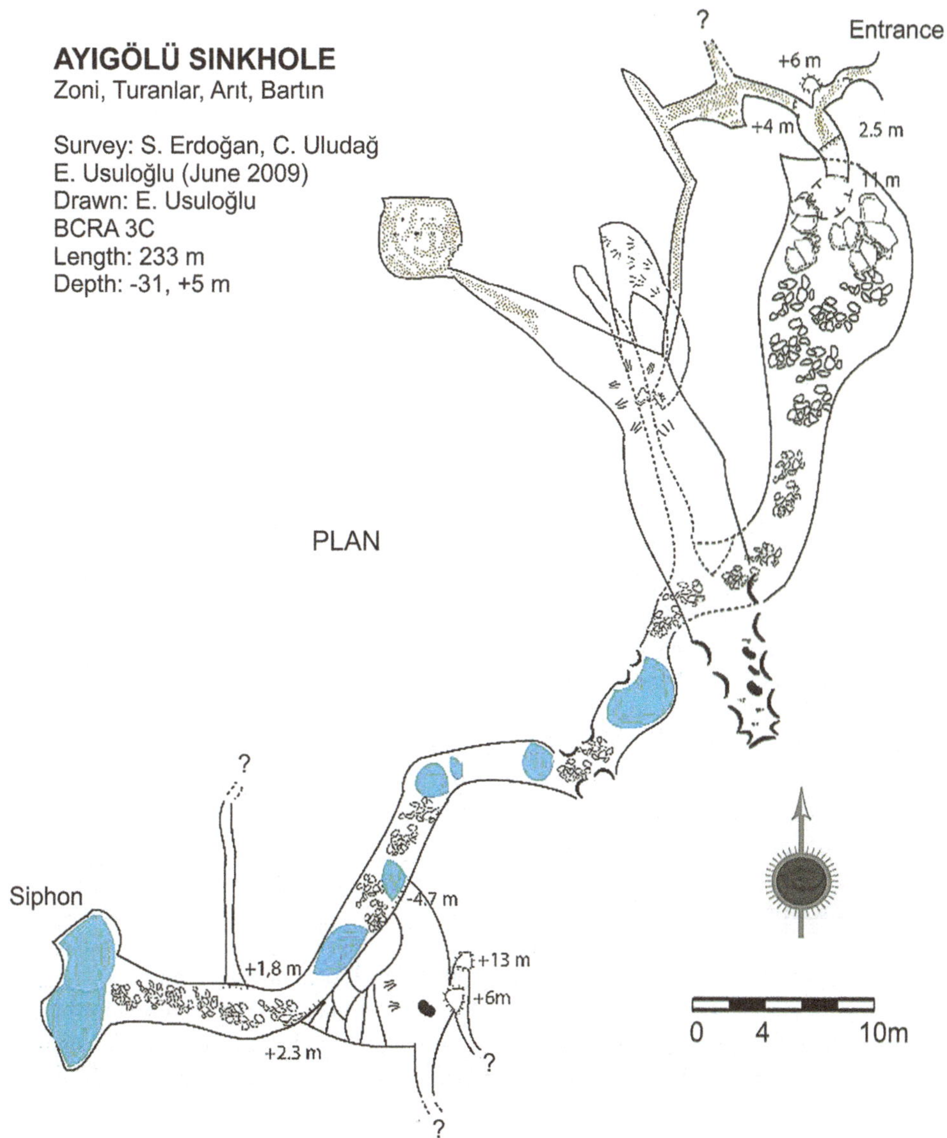

Fig. 3.9 Ayıgölü Sinkhole plan (drawn E. Usuloğlu)

Fig. 3.10 A passage in Ayıgölü Sinkhole (photo E. Usuloğlu)

logical survey by B. During and C. Glatz. The pottery discovered during the survey spans the Chalcolithic to the Classical Period.

The Cide-Şenpazar district contains cave sites that date back to 5500 BC and extend to 3000 BC. The data set for this period is distinct from that of the earlier Early Holocene period, for which data were only derived from exterior surfaces. The Ballıcı Cave site provides the most comprehensive dataset, followed by Okçular, Kılıçlı, Çadır, Gideros 1, and Derebağ caves.

Additionally, the cave contains pottery from the Classic Period, including examples of Byzantine pottery. Although only a small proportion of the pottery found in the cave is clearly of the Byzantine period, it is in very good condition and provides examples of kitchenware, probably from the Middle Byzantine period. One of the most notable aspects of the Byzantine religious landscape is the prevalence of hermit caves. These include the cave of Ballıcı Cave, where a small cache of Middle Byzantine coarse or cooking ware points to some form of habitation, possibly a hermit. Additionally, a small nail was discovered, which is likely of Byzantine origin (Düring & Glatz, 2015) (Figs. 3.11 and 3.12).

3.3.5 Buzluk Cave (No. 17)

The cave is situated in a forest approximately 30 min by foot from the slope on the ridge of Karafasıl Village, Sarma District. Buzluk Cave is regarded as one of the most aesthetically pleasing caves in the region. In this region, to the east and southeast of Pınarbaşı, there are pieces of covered limestone. This formation, comprising large, horizontal plates of Lower Cretaceous limestone, can be traced in places on the surface. However, these limestones are mostly covered with soil, and, above that, there is a dense forest. As a result of an old collapse at the entrance of this 483-m cave, some formations in this part are exposed, and a large part of the cave, which consists of a single, large hall, is illuminated. The cave contains a multitude of stalactites and stalagmites, some of which are of considerable size and have fallen to the ground, while others have been broken (Figs. 3.13 and 3.14). Additionally, there are cave formations in the small chambers at the entrance. The cave continues with another hall, which is reached by a high passage connecting to the main hall, and there are also very impressive formations.

On another excursion to the Buzluk Cave in April 2009, we encountered a remarkable discovery. Based on the cave's name (Buzluk = "icy" in Turkish), we anticipated a frigid environment and the potential for snow or ice. However, the fairytale-like image that greeted us surpassed our expectations. From the entrance of the cave, we observed a multitude of ice formations (Fig. 3.15). It is notable that no ice was present in the cave during August, indicating that this decoration is destroyed and remade by nature on an annual basis. Buzluk Cave's distinctive feature and impressive visual appeal make it a cave that merits further investigation.

The ice formation observed in this and similar exceptional caves must be explained in a manner that differs from the typical explanation. The ice within the cave is not in the form

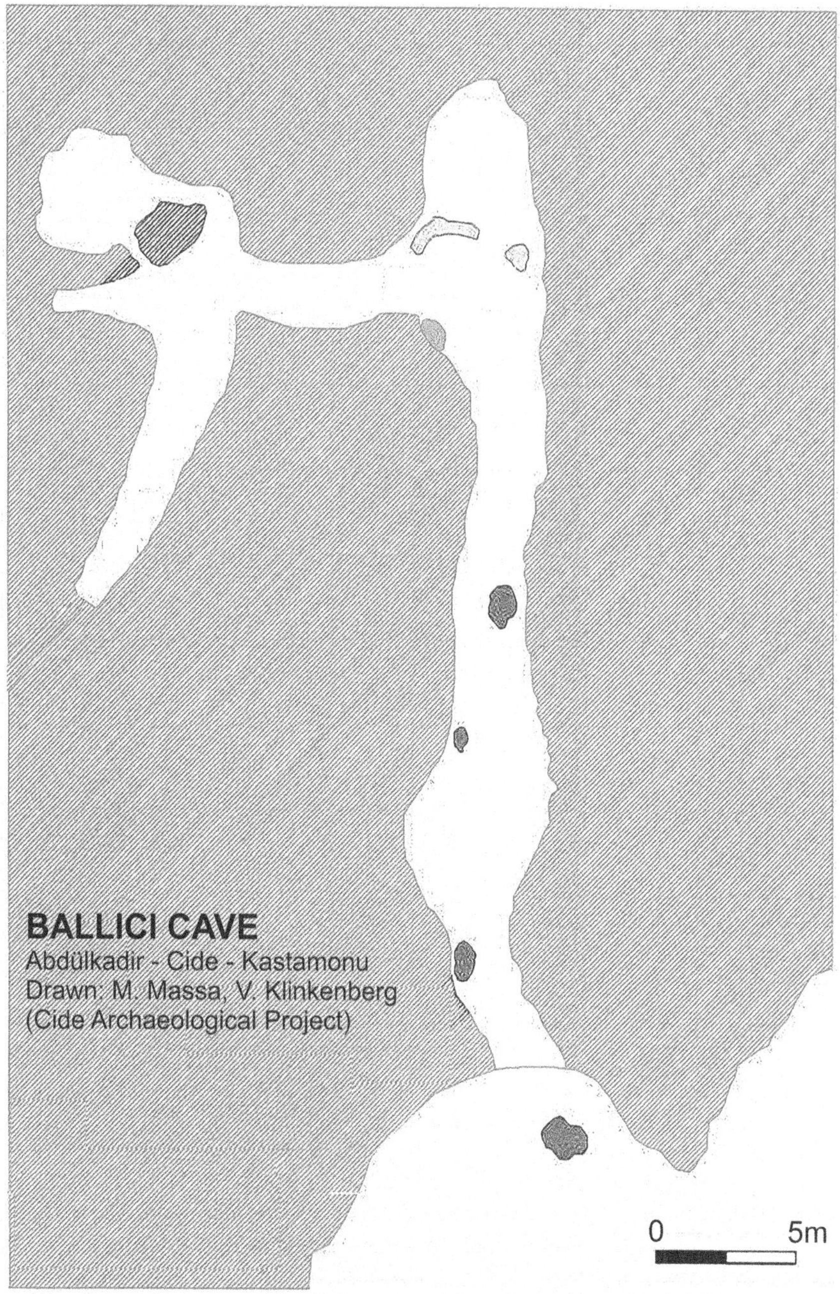

Fig. 3.11 Ballıcı Cave plan (drawn M. Massa and V. Klinkenberg)

of heaps, but rather in the form of formations such as stalactites, stalagmites, and travertines, which are spread throughout the cave. This indicates that the waters that originate from the surface in liquid form are frozen within the cave. In other words, it is necessary to explain the situation where there is a temperature below 0° within the cave, despite the presence of hot air on the surface.

As the caves have the average annual temperature of the surface on which they are located, in a typical cave at 1,250 m, temperatures below 0° are generally not expected. The fact that the temperature inside the cave dropped so low is probably due to the cold air coming in through the collapsed wide entrance. Thus, there is no typical cave environment, which is relatively isolated from surface conditions and has limited temperature changes throughout the year. During the cold winter months, the low temperatures on the surface cause the temperature inside the cave to drop below 0 °C, potentially reaching extremely low levels. Conversely, if we consider that the temperature on the surface will be below 0 °C during the same period, there will be no water reaching the cave from the surface. However, with the melting of snow in the spring, the optimal conditions for the emergence of the ice formations we are discussing are created. The rock mass's inertia ensures that the cave's tempera-

Fig. 3.12 Entrance of Ballıcı Cave (photo M. Massa)

ture will remain below 0 °C even after the surface has warmed. This allows for the emergence of ice formations when the snow melts on the surface, seeps into the cave through various cracks, and freezes again. Consequently, the most optimal period for observing the cave's beauty and formation is expected to be March to April.

It is reasonable to posit that this cave may be considered one of the three most beautiful caves in the region, along with the Ilgarini and Atak caves.

3.3.6 Büyükkaya Cave (No. 18)

The cave is situated on a rocky ridge above the village of Kayadibi in Ulus. It can be reached by a half-hour walk from the village. The cave was formed in Jurassic-Cretaceous-aged limestones along the NE-SW trending fracture plane. Büyükkaya Cave, which was suspended due to the vertical development of the valley and the resulting lowering of the karst floor level, remained entirely in the vadose zone. The total length of the cave is 113 m. Büyükkaya Cave, which has a wide and high entrance that can be seen from the main road, consists of a single gallery that narrows as it progresses. In the inner parts of the cave, which measures 6 × 6 m at the entrance, these values decrease to 1 × 1 m. The cave then expands again towards the last sections, narrows again, and ends (Figs. 3.16 and 3.17).

The cave, whose floor is covered with small blocks and clays in the entrance section, is rich in wall and cover flowstone and dripstone formations, especially in the last sections, as it progresses to the inner sections. The cave, which reaches a height of 16 m above the entrance, is part of a fossilized system that has completed its development, similar to other caves in the region. Büyükkaya Cave, which is devoid

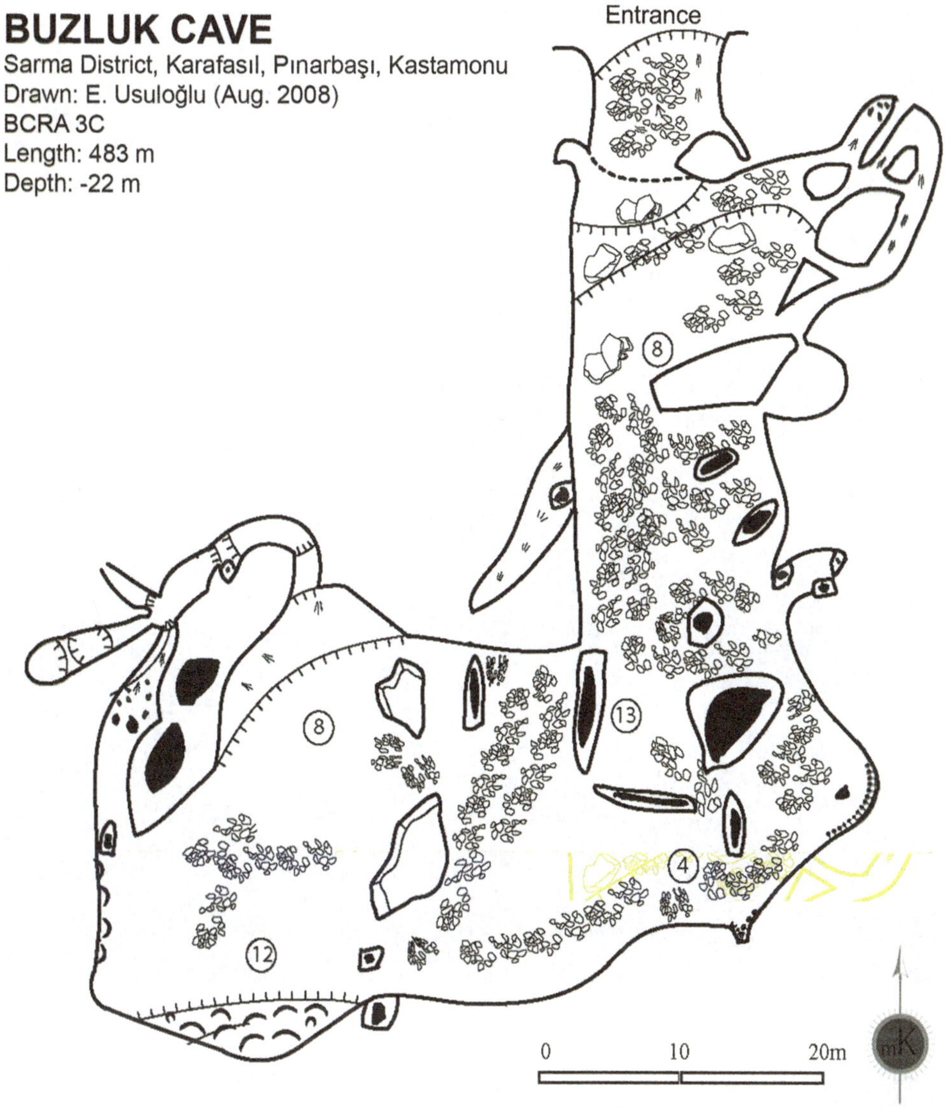

Fig. 3.13 Buzluk Cave plan (drawn E. Usuloğlu)

of groundwater activity, is completely dry, with the exception of small muddy areas formed by the infiltration of precipitation-derived waters. The cave was explored by MTA in 2002, and according to the report, no significant living organisms were observed within the cave (Nazik & Törk, 2003b).

3.3.7 Çadır Cave (Fındık Cave) (No. 19)

It is located 20 m above Gideros from Bartın Kurucaşile's main road. It's a small, horizontal cave with a minimal formation that may have served as a barn for a very long time. It was documented in 2010 as part of the Cide Archaeological Project (Figs. 3.18 and 3.19).

Outlying structures and architectural remnants that could be seen, if not fully articulated, in a survey, as well as a renovated natural cave located directly below the church, all support the idea that this location once served as a monastic center. Although carved ledges, benches, and wall sockets were added at some point while it was being used with the church, Çadır Cave was a sizable cave that was used for much longer than the Byzantine period. Its proximity to the church—which is situated atop it—suggests that the two may be related. Once more, the ceramics were unremarkable, but one tiny fragment of glazed pottery indicates that the cave was used in the Late Byzantine era.

Throughout the Byzantine era, the church was altered and shrunk in size to make room for a grave, possibly for a revered local figure. A portion of Çadır Cave, which was included in the church complex, was built above it. Without

Fig. 3.14 The first chamber of Buzluk Cave (photo G. Ambrus)

Fig. 3.15 Ice stalactites in Buzluk Cave (photo A. Yamaç)

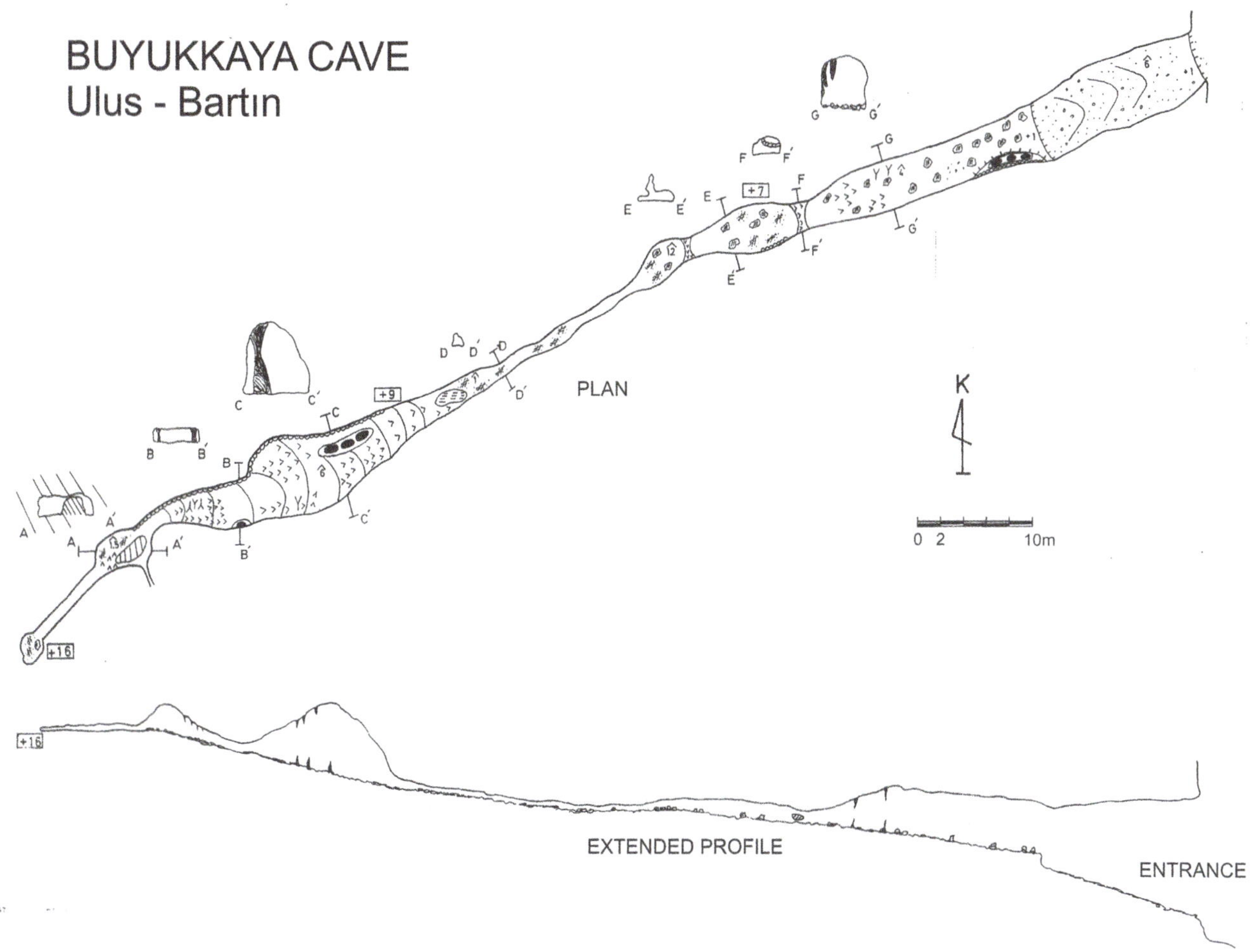

Fig. 3.16 Büyükkaya Cave plan and extended profile (Nazik & Törk, 2003b)

Fig. 3.17 Entrance of Büyükkaya Cave (photo K. Törk)

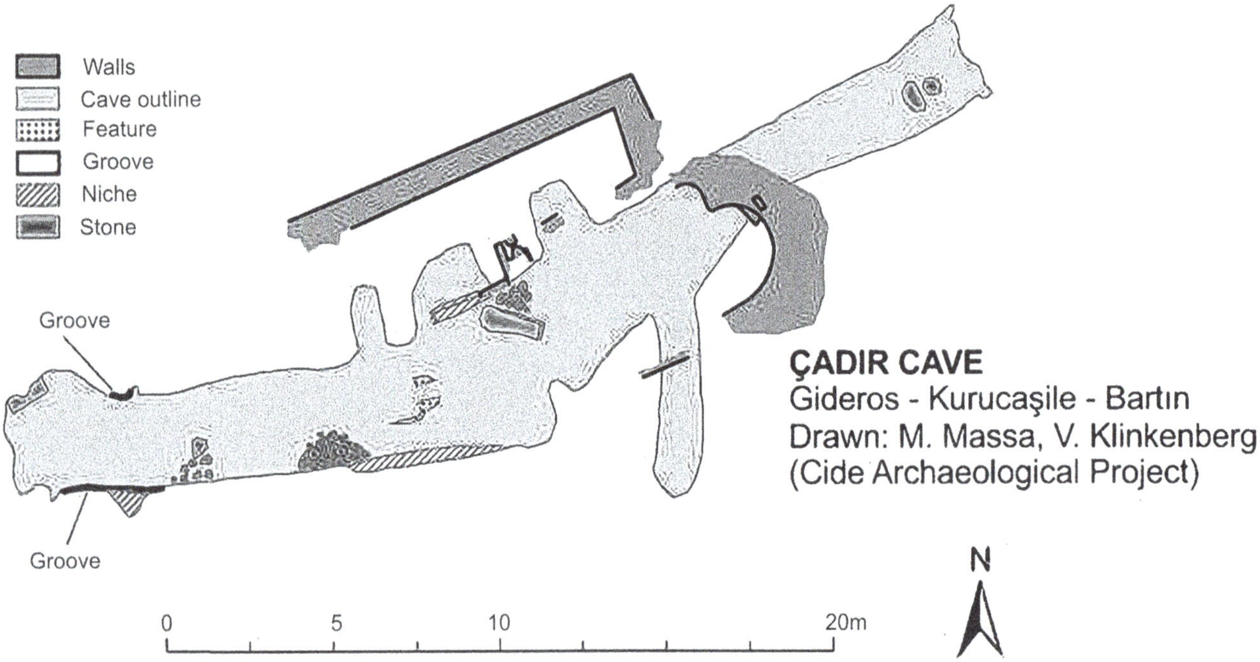

Fig. 3.18 Çadır Cave plan and extended profile (drawn M. Massa and V. Klinkenberg)

Fig. 3.19 Entrance of Çadır Cave (photo M. Massa)

a doubt, the connection between the cave and the church is important. Although a practical purpose, like cool storage, is conceivable, it's also possible that the object serves a culturally or spiritually significant purpose. A prehistoric pottery fragment and pieces of human bone were discovered during our exploration of the cave, which may indicate the long-term cultural significance of this area.

The monastic community would have likely owned land in the Abdulkadir Valley, and its agricultural output would have provided financial support for the institution and freed the majority of its members from daily subsistence tasks, allowing them to devote more time to their religious pursuits. This was a pilgrimage site connected to an unnamed holy figure given the size of the complex, the existence of the cave, and the proximity of Gideros Cave (Düring & Glatz, 2015).

It is known that numerous caves in Gideros remain unexplored due to the dense forest cover that extends from the road up to the mountains. Conversely, the dense vegetation and inadequate description also led to the rediscovery of the same caves. Although the maps in question exhibit slight discrepancies, it is possible that the cave designated as "Fındık Cave," which was explored by LUSS in 1990, may be identified as "Çadır Cave." Furthermore, another analogous double exploration is presented in Sect. 3.3.43.

The LUSS report provides a brief description of "Fındık Cave" in a single paragraph:

> A large entrance crawl after two meters enters a large passage, then a wide, low chamber strewed with calcified snail shells. The climb up the calcite ramp at the end of the chamber leads to the "Dormouse Grotto" and further passage for 80 m until all the ways are blocked by sandy infill in low areas. The cave system appears very impressive and is well decorated. The large phreatic passage throughout the cave is largely sealed up, and digging might well extend the cave further. Our guide told us that entrances to more caves were found in the hillsides above the entrance to Fındık Magara, but we couldn't locate any of these (Holland, 1991).

3.3.8 Çengelkuyu Sinkhole (No. 22)

Çengelkuyu Sinkhole is on the plateau of Kerte Village, which is approximately 4 km north of Ilıca—Pınarbaşı. The cave was explored by BÜMAK and published in the preliminary report in Delta 7 (Kuruçayırlı & Osmanoğlu, 2004).

According to the report, this sinkhole, which was surveyed in October 1998 and is 199 m deep, also has an undiscovered branch. A siphon marks the end of the cave (Fig. 3.20). As one of the deepest caves in the area, Çengelkuyu Sinkhole deserves a serious and thorough investigation. All that is known about it is a very basic report consisting of a few lines.

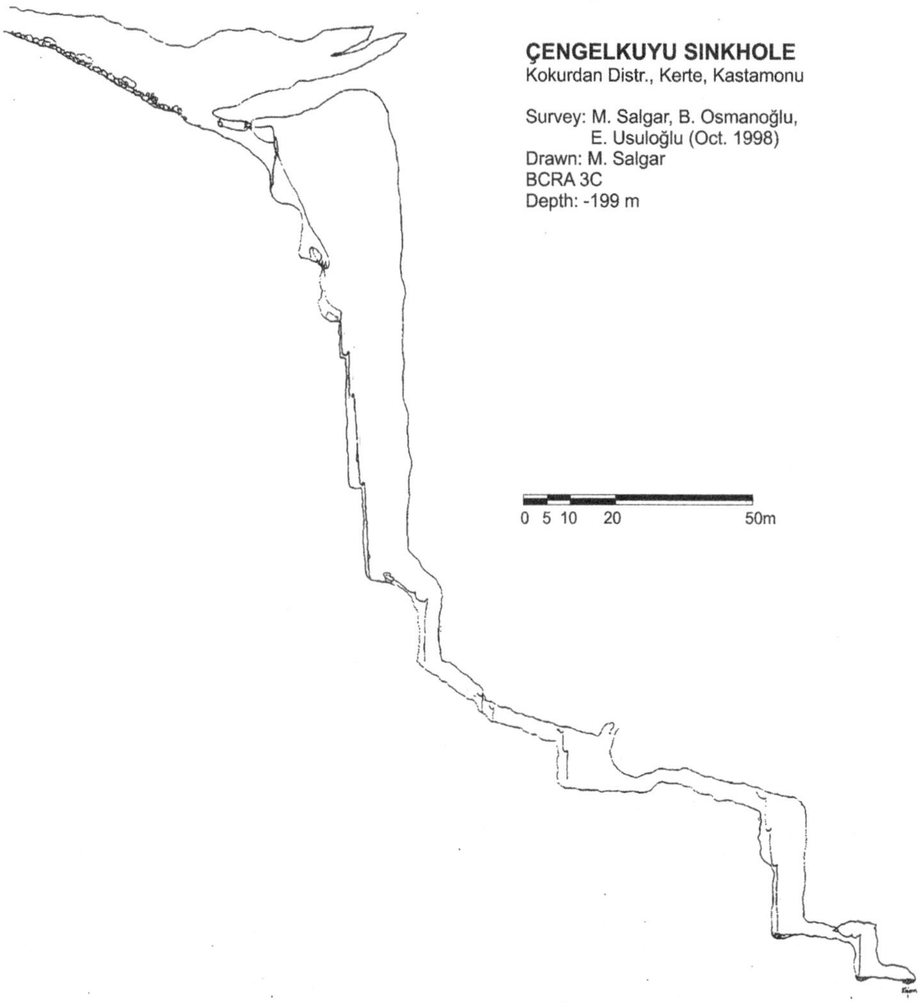

Fig. 3.20 Çengelkuyu Sinkhole extended profile (drawn M. Salgar)

3.3.9 Çıkrıkkapı Sinkhole (No. 23)

Çıkrıkkapı is located north of Sümenler Village in Pınarbaşı. It is situated between the Sorkun and Kapaklı sinkholes. Çıkrıkkapı is situated on the band where the impermeable layer sinks under the limestone block. In this zone, which extends almost along the entire southern line of the plateau, the streams of all basins have formed separate sinkholes. A total of 18 sinkholes of varying depths are distributed in a linear fashion in this region, with Çıkrıkkapı representing one such example (Fig. 3.34).

Çıkrıkkapı is a 350-m-long, −166-m-deep active sinkhole that collects the water of a small basin. The sinkhole features a relatively short main axis measuring approximately 80 m, numerous short descents in a north-south direction, and culminates in a siphon that is partially obstructed with mud. The cave was first explored by BÜMAK in 1987 (Albukrek, 1993a) (Fig. 3.21).

3.3.10 Çıngırık Sinkhole (No. 25)

The cave is situated at the base of the limestone wall, 200 m prior to Kokurdan, in the southern region of Okçular Village of Cide. Çıngırık is an active sinkhole that collects the water of a very large basin. The remains of a water mill located in front of the cave, which is no longer in use, provide insight into the high flow rate of the stream discharging the water of this basin into Çıngırık Sinkhole during the high season.

The cave's entrance, which is formed in solid Cretaceous limestone, measures 12 m in diameter and 15 m in depth, making it a notable feature (Fig. 3.22). A sloping gallery and a brief descent of approximately 6 m follow the initial chamber. Subsequently, two additional descents are encountered, the first of which is particularly challenging due to the presence of thick mud. This descent extends for 50 m, and the second descent is 25 m in length. The resulting chamber is notable for its impressive 25-m ceiling. The stream reaches a sump at this point. The fault fracture that occurs

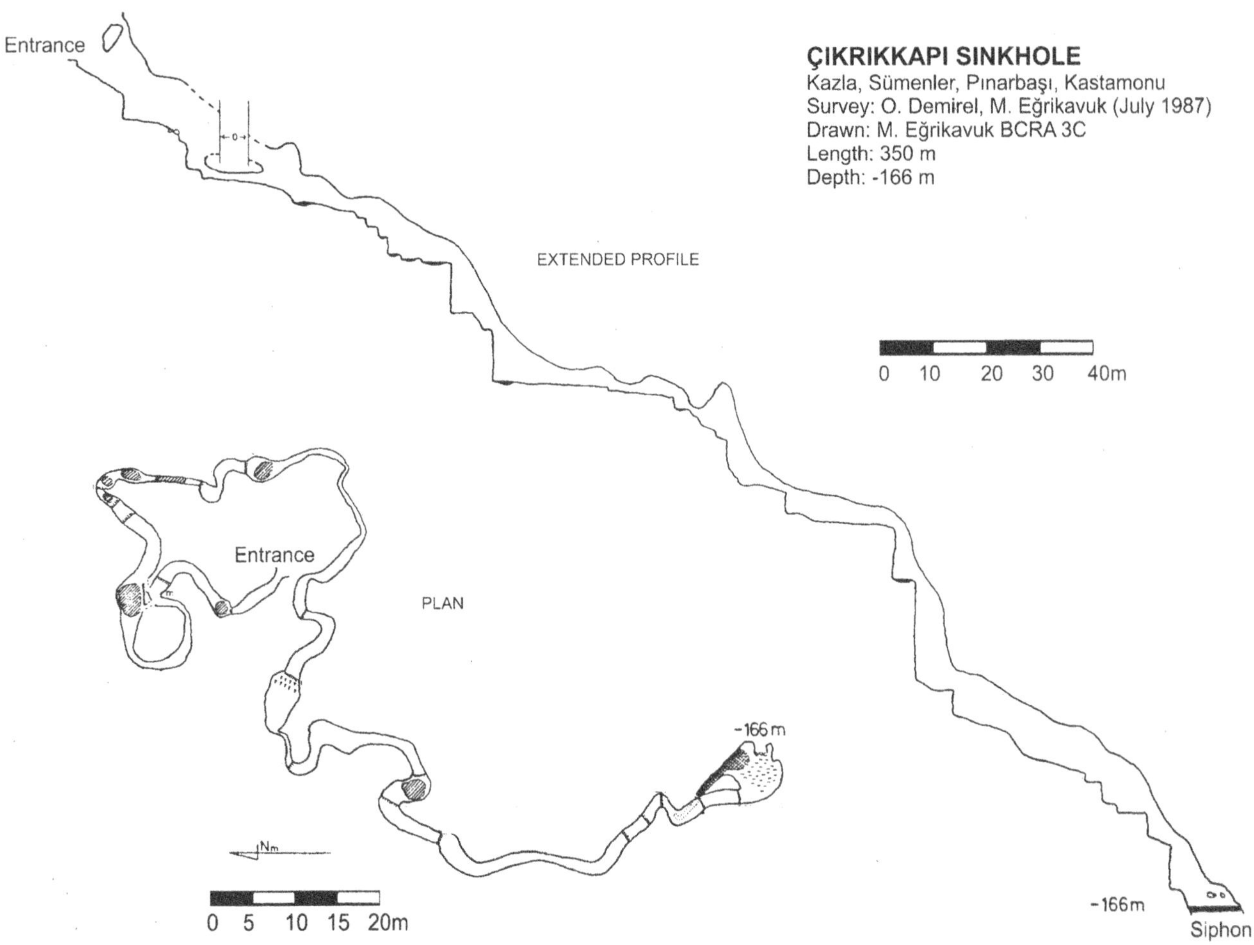

Fig. 3.21 Çıkrıkkapı Sinkhole plan and extended profile (drawn M. Eğrikavuk)

Fig. 3.22 First descent of Çıngırık Sinkhole (photo A. Yamaç)

within the cave has such a precise inclination that, even at the deepest point, light from the entrance can be seen (Fig. 3.23).

Despite its 93-m depth, Çıngırık Sinkhole is only 150 m above sea level and only 3 km away from the Black Sea. It is our contention that the stream should continue to flow horizontally after the siphon at the bottom of the cave and drain into the Black Sea.

3.3.11 Dağlı Sinkhole (No. 32)

The Dağlı Sinkhole is situated in a forest approximately 3 km from Dağlı Village, situated between Şenpazar and Cide. It is one of the most attractive caves in the region, distinguished by its expansive, elliptical opening within a dense vegetative setting, measuring approximately 80 by 40 m (Fig. 3.24). This geological phenomenon represents a remarkable example of the natural world. In essence, it resembles a vast crater within the forest, akin to an immense meteorite pit. The cave system is comprised of four sections: the Main Branch, the Side Branch, the Cık Hole, and the Watery Branch. In addition to the Main Branch, the Cık Hole has been identified as the deepest known cave in the Black Sea region, with a depth of −279 m. The entrance of the Main Branch descends 80 m as a single shaft, after which the cave extends downward in two main branches. The branch on the right side, according to the entrance, is particularly challenging to access. The branch to the left of the entrance extends up to −180 m in the form of a wide shaft that is independent of the horizontal connection indicated on the map.

The Watery Branch, a waterfall situated just below the zero point, joins the main branch at a depth of −45 m within the cave. The opening on the opposite side, located 5 m after the beginning of the waterfall, is designated as the Cık Hole. This hole, which was initially believed to be a horizontal cave during the inaugural expedition, has since become the deepest, longest, and most significant component of the cave

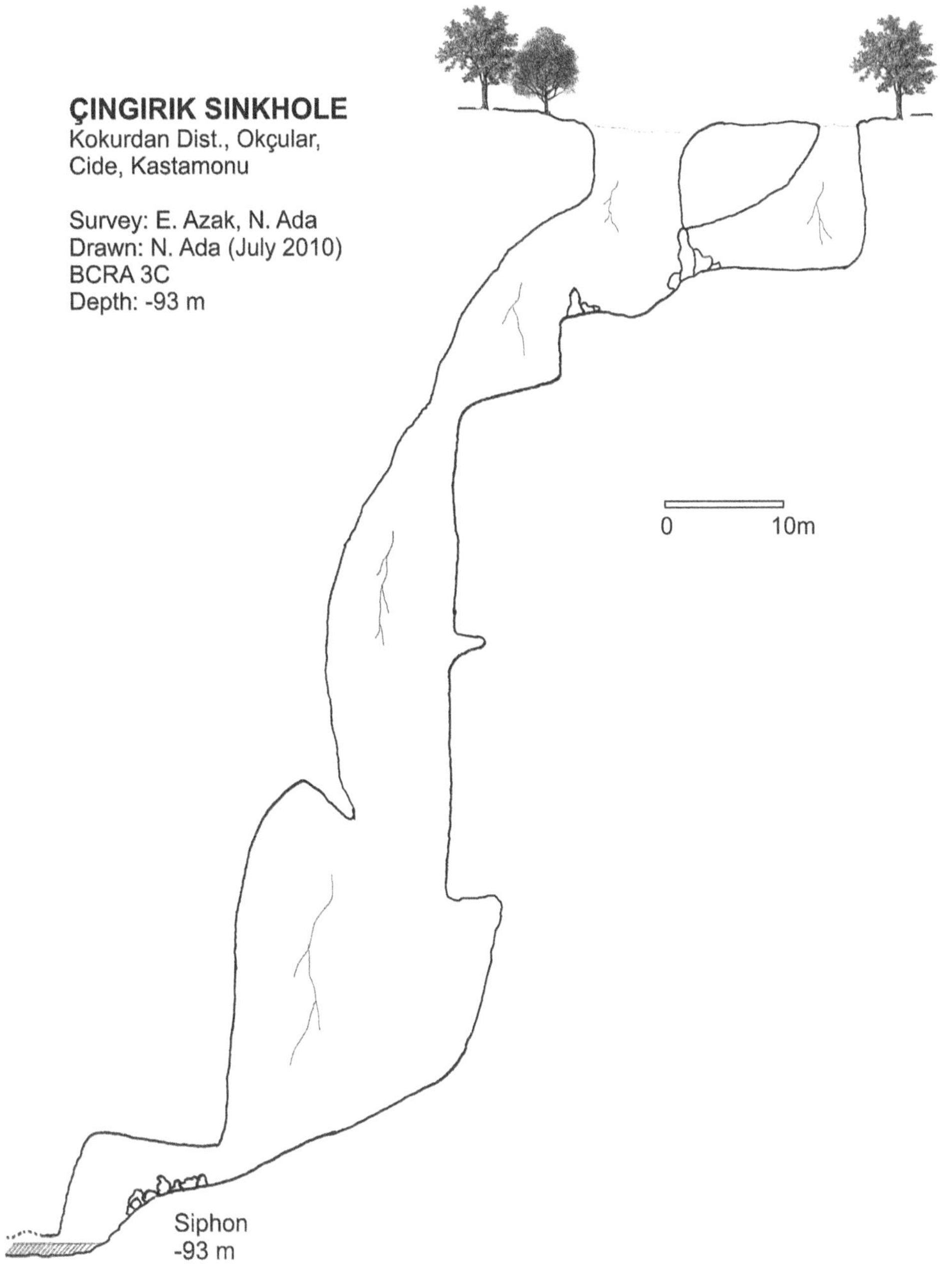

Fig. 3.23 Çıngırık Sinkhole profile (drawn N. Ada)

system. Cık Hole, situated on the opposing wall of the cave at the point where the waterfall commences, descends independently from the principal branch to a depth of −279 m, exhibiting a fossilized and highly muddy composition (Fig. 3.25).

Cık Hole, which exhibits the characteristics of a sinkhole and was likely active during the geologic period, has become fossilized as a result of tectonic movements over time and has been displaced from the waters of the stream that feeds it. The cave is devoid of any significant water source, with the exception of a minor seepage from the ceiling. Notwithstanding the fact that the water level in Watery Branch rises with the precipitation, no water enters the cave. The course of Cık Hole begins with a horizontal trajectory characterized by numerous short descents.

The final section of the horizontal track is a 10–15 m horizontal corridor designated as the "Dance Floor." The floor's resemblance to a completely smooth marble and its slipperiness have led to the name "Dance Floor." It is estimated that this formation has survived from the period when the cave was active to the present day due to the formation of a protective layer of mud on it. The vertical track, which begins at the end of the aforementioned track in the form of a cliff, descends approximately 50 m, prompting the visitor to

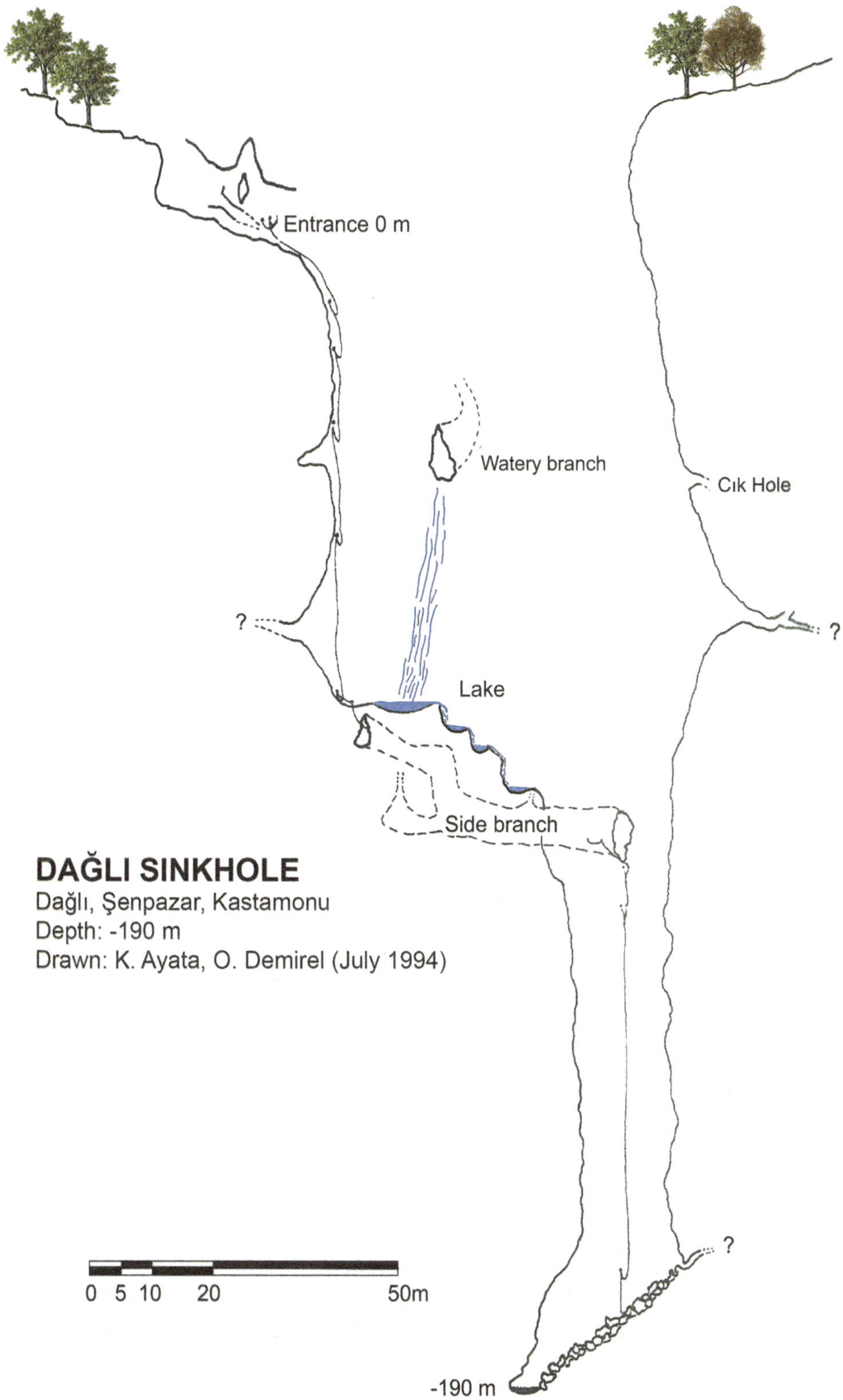

Fig. 3.24 Dağlı Sinkhole profile (drawn K. Ayata, O. Demirel)

believe that they are in a sinkhole. The cave continues vertically and opens into a hall. Mud is carried into this hall by a crack connected from the opposite side of the descent. The other point reached by the long descent following the hall is a second hall.

Conversely, the Main Branch, which exhibits the characteristics of a typical collapse doline, has become active over time (Figs. 3.26 and 3.27). There are several hypotheses regarding the formation of the Dağlı System. The Cık Hole, which collected the water of the basin, which could be considered quite small in its earliest times, was cut off from the

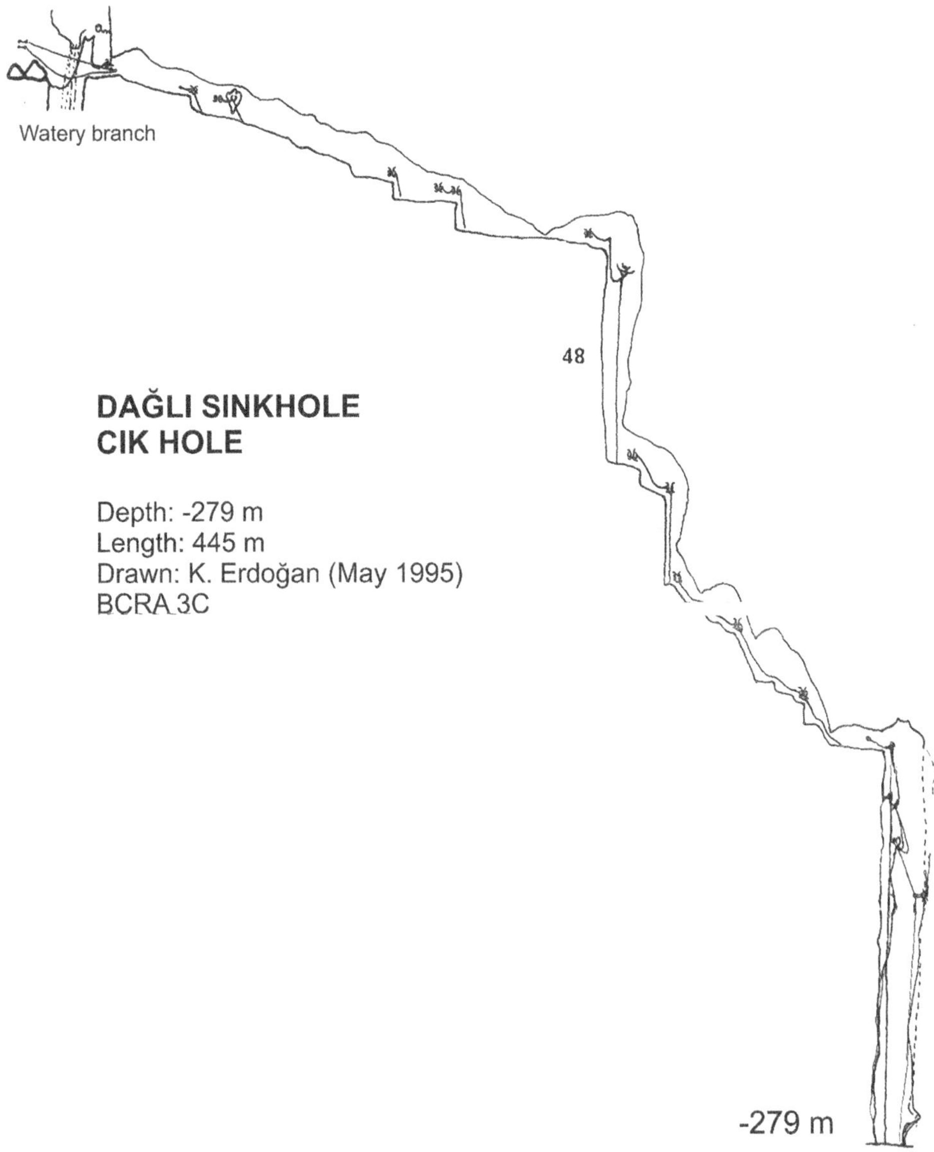

Fig. 3.25 Cık Hole profile (drawn K. Erdoğan)

waters of the stream that fed it as a result of tectonic movements. This resulted in the formation of the Water Branch, which eroded to deliver its water to the main tributary. The Main Branch is a collapsed doline. The primary cause of these depressions is the alteration of the underlying geology by erosion and groundwater transport. Additionally, the width of the volume indicates the presence of a substantial underground water basin. The sluice-like formation at the end of the Main Branch supports this assumption. The Dağlı Sinkhole was explored by BÜMAK between 1993 and 1995 and published in Delta 7 (Kuruçayırlı & Osmanoğlu, 2004).

3.3.12 Derebağ Village Cave (No. 33)

Derebağ Village Cave is a significant cave (approximately 220 m in length) situated at the foot of a relatively steep mountain above the village of Derebağ in the easternmost region of the Cide district (Fig. 3.28). The cave was subjected to extensive research and published as part of the Cide Archaeological Project.

Although it does have some narrower passages and descends and ascends significantly, the cave is generally linear and has a consistent width and height. The cave is characterized by the presence of numerous collapsed stones that have fallen from the roof. Pottery was recovered from a sedimentary deposit beneath the collapsed rocks, which are dated

Fig. 3.26 Dağlı Sinkhole Watery Branch with a caver at the bottom (photo D. Lee)

to various historical periods. The only cave site that does not display signs of looting is Derebağ Village Cave. This serves to highlight the remarkable discovery of numerous sherds from a variety of eras, including the Chalcolithic, Early Bronze Age, Early Second Millennium BC, Iron Age, and Byzantine periods. The cave floor is currently covered in stone that has fallen from the roof, but it is unclear if the cave was always this unstable.

A considerable quantity of dark gray pottery sherds from handmade, insufficiently fired vessels was discovered in the Derebağ Village Cave. Initially, all of the sherds were assigned a Chalcolithic age, but some of their characteristics did not readily fit with this classification. One notable distinction between the sherds and the assemblage from Okçular Cave is that a significant proportion of the sherds contained plant temper. Moreover, a significant proportion of the vessels were manufactured using rudimentary techniques, with surfaces that had been inadequately treated and pastes that had been incorrectly sorted. Additionally, the vessels exhibited large inclusions and uneven wall thicknesses.

The pottery is distinct from typical Chalcolithic pottery in the Cide-Şenpazar region and elsewhere in all these ways. It is made with care and is nicely smoothed and burnished, even if it is fired inefficiently.

All of these formal types were discovered together in the Derebağ Village Cave.

One of the most distinctive features of the pottery is the dark brownish horizontal handle with a black mineral temper that resembles the volcanic sand of the Sinop region. Another brown fragment features a horned handle with a protruding tip on the carination. Horned handles are known from Greece and Italy during the Late Bronze Age and the Early Iron Age. This example may indicate contact, although it is difficult to draw exact parallels with the West (Düring & Glatz, 2015).

Fig. 3.27 Dağlı Sinkhole main shaft (photo D. Lee)

Fig. 3.28 Derebağ Village Cave entrance (photo Cide Archaeological Project)

3.3.13 Döngelyanı Sinkhole (No.36)

The cave is situated approximately 500 m to the east of Topmeydanı in the northern region of Sümenler Village of Pınarbaşı. Although the Döngelyanı Sinkhole drains a relatively small basin, it has a high flow rate during the rainy season.

BÜMAK undertook an exploration and measurement of the Döngelyanı Sinkhole in its 1985 survey, with the exception of a minor tributary that rose 30 m above the entrance. Approximately 20 m after the entrance, the first long descent, measuring 48 m in length, is encountered. The majority of this descent is open, with no obstructions. Subsequent to the pool situated below the aforementioned descent, the cave exhibits a series of more constrained passages that warrant further investigation. The ceiling begins to descend after three brief descents in the gallery, which is covered in crumbled rocks.

After 30 m of horizontal advancement, a branching off to the right reaches the final descent of 24 m. A lake with silty clay shores can be found in the small area. The gallery then divides into several branches, which take the form of tunnels. In the other branch, which is to the left, the ceiling also descends further. The tunnels have a diameter of 1 m, and the ground inside is covered with clay alluvium. The tunnels, which range in length from 15 to 20 m, split into new branches and become increasingly narrow until they are impassable. The discovery of green leaves on the ceilings of the galleries indicated that they had been completely flooded during the rainy season. The cave ends at −195 m with a siphon blocked by alluvium. The total length was measured as 530 m. During the surveys conducted by ASPEG in 2008, many small species were found in the cave (Albukrek, 1986; Yamaç, 2008) (Figs. 3.29 and 3.30).

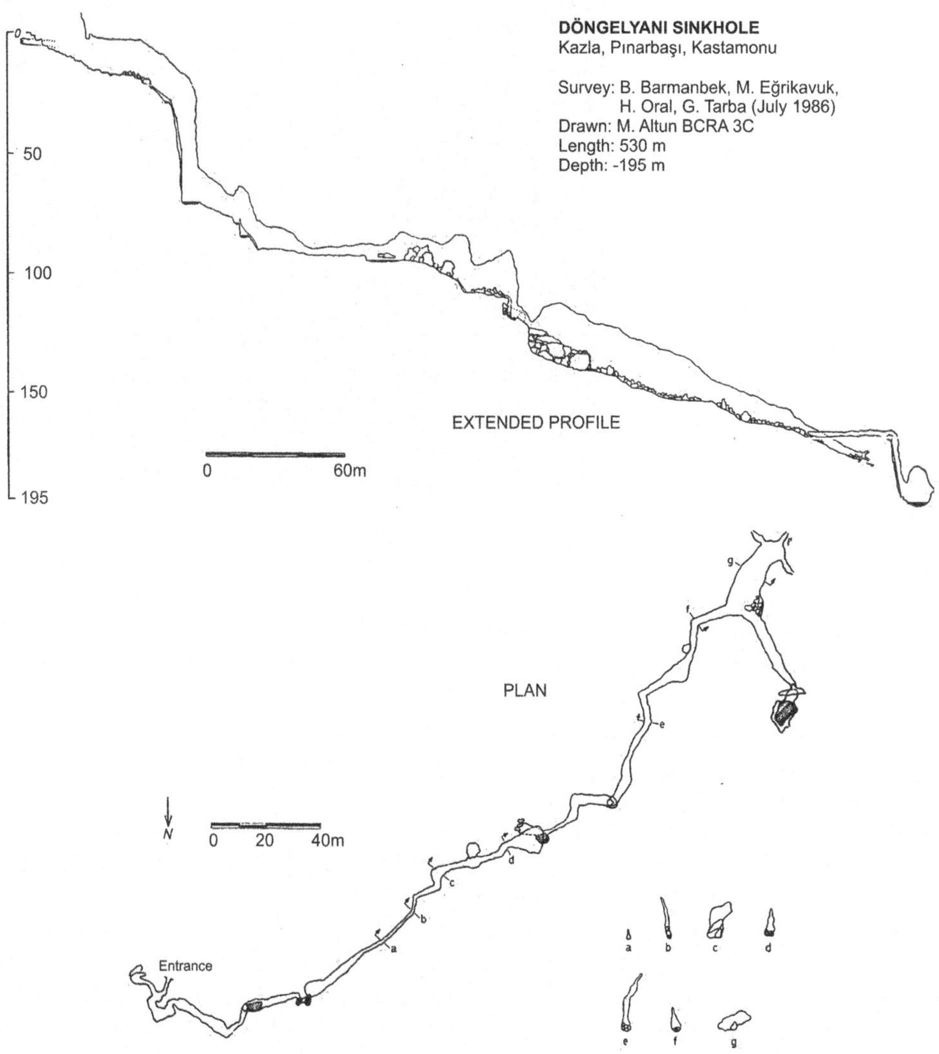

Fig. 3.29 Döngelyanı Sinkhole extended profile and plan (drawn M. Altun)

Fig. 3.30 Entrance of Döngelyanı Sinkhole (photo M. Albukrek)

3.3.14 Ejder Cave (No.38)

The Ejder Cave is arguably one of the most visually striking and impressive caves in the Küre Mountains National Park. It is situated in the northern region of Yamanlar Village in Pınarbaşı, on the trail within the forest leading to Ilgarini Cave. Covering an entire rocky hill, this cave descends to a depth of −58 m with two particularly large openings. The descent of 58 m in such a large cave is a visual feast (Figs. 3.31 and 3.32). After this descent, the cave continues with an inclined ledge and reaches a depth of −75 m. Due to the temperature difference from the ground, fog occurs at the entrance of the cave from time to time. It is known that this cave, which was probably formed as a result of active erosion, receives water during rainy periods. Also known as "Isputluk Sinkhole" by the local population, the cave was separately explored by MAD and BÜMAK, with the results published in a brief note in Delta 7 (Fig. 3.33). During the research conducted by ASPEG, numerous small living species were identified within the cave, with the results of the sampling published in OBRUK 1. The cave is notable for its high biodiversity, containing endemic species from the Diplopoda, Chilopoda, and Pseudoscorpionida groups (Kuruçayırlı, 2004; Yamaç, 2008).

3.3.15 Eşek Çukuru Sinkhole (No.42)

It is not only foreign cavers who make an irreparable mistake when naming a newly discovered cave; local cavers also make mistakes from time to time. This cave is an example.

Eşek Çukuru Basin is located to the west of the Sorkun Basin, north of the Yamanlar Village of Pınarbaşı. It is the largest basin in the region after Sorkun. Another small basin

Fig. 3.31 Entrance of Ejder Cave (photo G. Ambrus)

Fig. 3.32 Entrance of Ejder Cave (photo M. Renda)

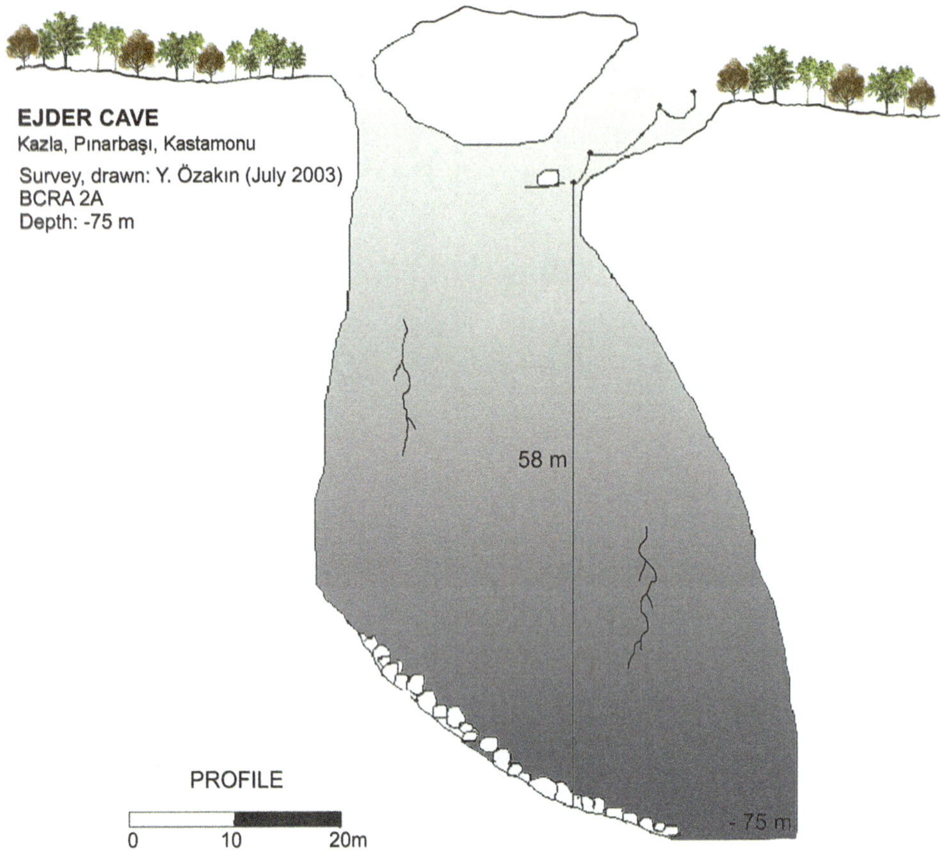

Fig. 3.33 Ejder Cave profile (drawn Y. Özakın)

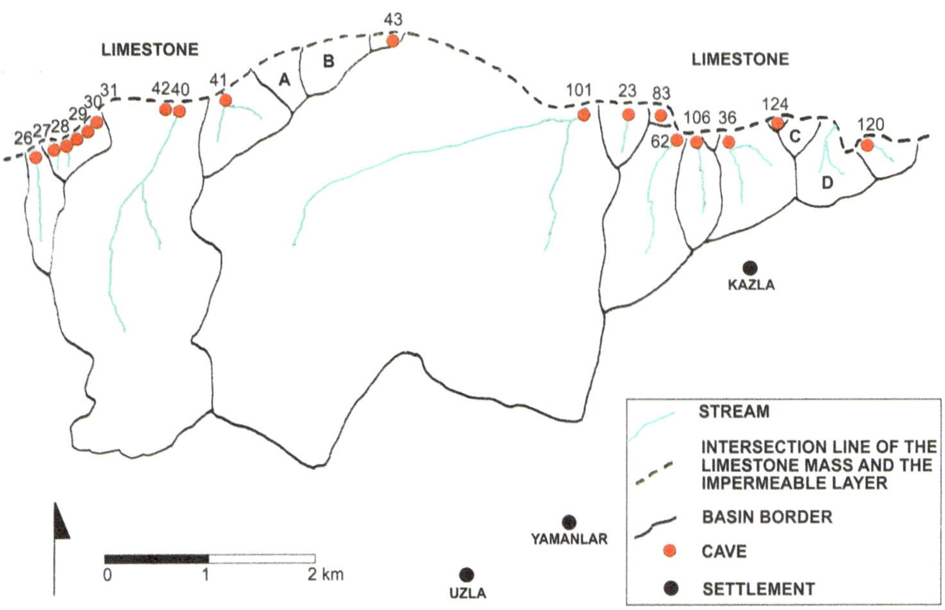

Fig. 3.34 Map showing the basins and the impermeable layer (Ulus Formation)—limestone (İnaltı Formation) border. Cave numbers are from Fig. 3.2. (**a**) This basin has a blocked sinkhole; (**b**) This basin has a collapsed sinkhole; (**c**) This basin has no sinkhole; (**d**) This basin has a collapsed sinkhole. Simplified from Albukrek (1993b)

just west of this is named 'Çovurma Tepe' by BÜMAK about a nearby hill. Six caves that were discovered in this basin during the 1980s were named 'Çovurma Tepe' (No. 26–31). During the same period, two more caves were discovered in the Eşek Çukuru Basin, east of Çovurma Tepe and west of Sorkun, and named Eşek Çukuru Cave No. 1 and 2 (No. 40 and 41) (Albukrek, 1993b) (Fig. 3.34).

In 2009, another cave was discovered in the same area, designated "Eşek Çukuru 2 (Çovurma Tepe 6) Cave." This designation is confusing, as it is identical to that of another cave and also shares the same name as Çovurma Tepe, which is situated approximately 500 m to the west. In order to avoid any further confusion, the name of this cave has been changed to "Eşek Çukuru Sinkhole."

The cave is situated to the north and in close proximity to Eşek Çukuru Cave 1. Eşek Çukuru Sinkhole is an active cave with six different entrances, the majority of which are obstructed by depressions or mud. Conversely, some of these entrance passages were suspended in time, resulting in the formation of a vadose zone in the side branches. The two exceptions are the branch that turns east and siphons immediately after the entrance and the main gallery of the cave. The main gallery proceeds in a northerly direction for a considerable distance, almost in a straight line, and at the end divides into two branches, which then end with siphons. Local pools are present along the main gallery, though none of them are particularly deep. However, the traces on the gallery walls indicate that the cave was a significant source of water during the active periods.

With a length of 1231 m and a depth of 29 m, it is one of the few horizontal caves in the region (Figs. 3.35 and 3.36).

3.3.16 Gökdeli (Mızrak—Buz) Cave (No. 44)

This time, the cave has three names. During our research in 2008, we were informed that the cave was known as "Gökdeli Cave." In 1992, the British FACC team designated the cave as "Mızrak Cave," a name derived from the discovery of a bronze spear during their research (Mızrak = "Spear" in Turkish). In Ahmet Ünal's article on the spear, it is written that the local population in the vicinity of Pınarbaşı refers to the cave as "Buz Cave" (Ünal, 1999).

The cave is situated in the vicinity of Güvercinlik Village in the Pınarbaşı region. In contrast to the formations observed further north within the national park and its periphery, the region in the south, encompassing Pınarbaşı, is Cretaceous-aged and largely comprised of covered limestone. The surface of these covered limestone deposits is characterized by the absence of horizontal caves or sinkholes. Gökdeli Cave,

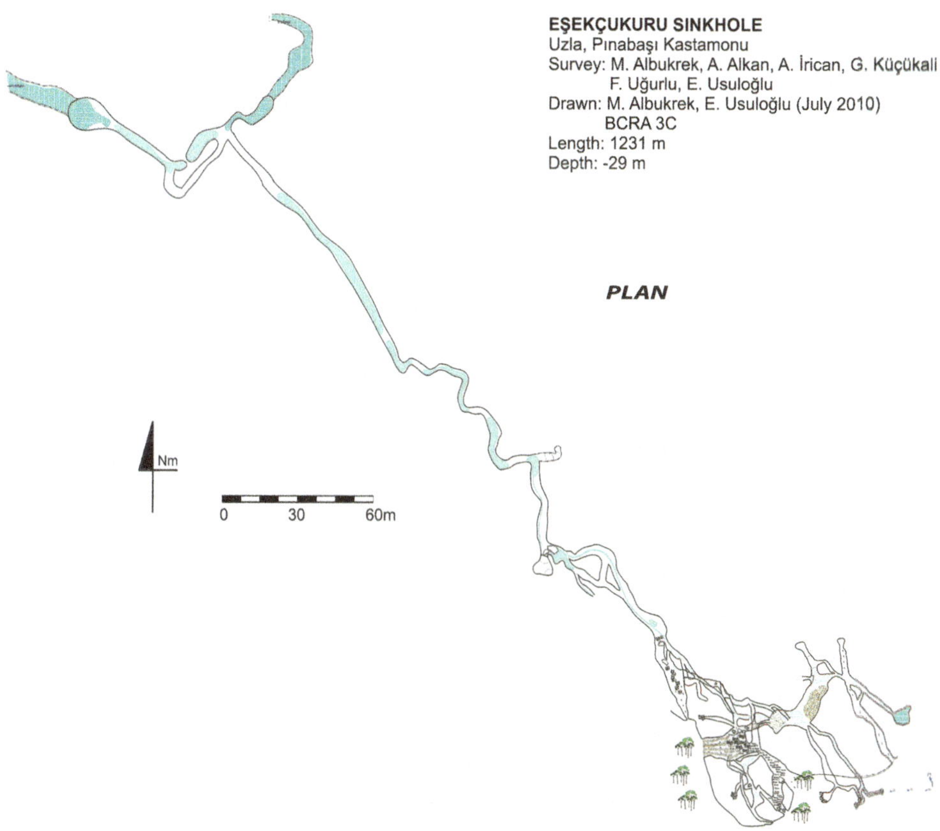

Fig. 3.35 Eşek Çukuru Sinkhole plan (drawn M. Albukrek, E. Usuloğlu)

Fig. 3.36 Entrance of Eşek Çukuru Sinkhole (photo M. Albukrek)

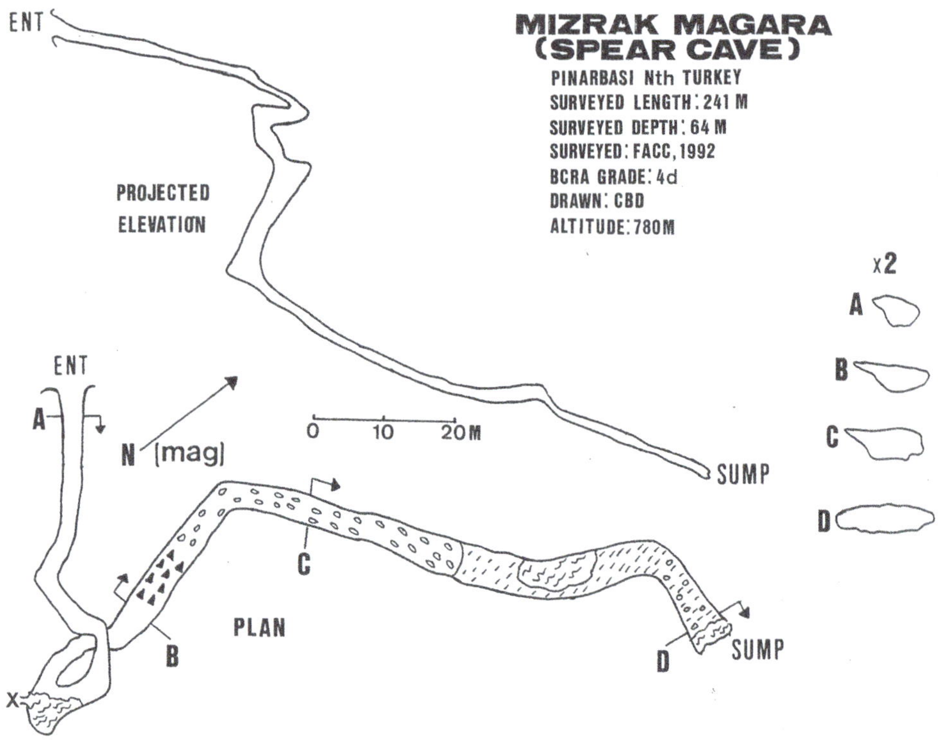

Fig. 3.37 Gökdeli (Mızrak) Cave plan and extended profile (drawn FACC)

Fig. 3.38 Gökdeli (Mızrak) Cave, the entrance (photo Ç. Çankırılı)

with a total length of 241 m and a depth of −64 m, is one of these sinkholes (Figs. 3.37 and 3.38). The cave comprises two steep descents and one phreatic passage, which descends to a static sump that appears to continue downwards. A brief extension was subsequently investigated at the initial pool, which is characterized by a narrow, upward-directed passageway lined with flowstone. This leads to a brief ascent, culminating in a keyhole, which in turn opens into a sump (Harris, 1992).

3.3.17 Gökpınar Cave (No.45)

Gökpınar Cave is situated on the periphery of a field in Makaracı Village of Amasra, Bartın. The cave is accessed by descending 3 m through a small and narrow hole in front of a limestone wall.

Initially, it was assumed that Gökpınar was a fossil cave due to the lack of an apparent stream and the fact that the area where the cave is located is entirely covered in soil. However, approximately 10 m after the entrance, it became evident that the cave was phreatic. A small stream was observed flowing inside the cave in a south-north direction just after the entrance. This stream was observed to have formed deep ponds in several places. At the conclusion of this 282-m-long cave, it was determined that the stream in question originates from another siphon (Fig. 3.39).

It is likely that this water is emerging from Kocareis Cave (No. 74), situated approximately 400 m to the south of Gökpınar Cave and aligned in the same direction. In contrast, a few minor springs were identified on the cave walls. The cave exhibits a considerable degree of flowstone formation and displays a high degree of biodiversity in its dry sections (Fig. 3.40).

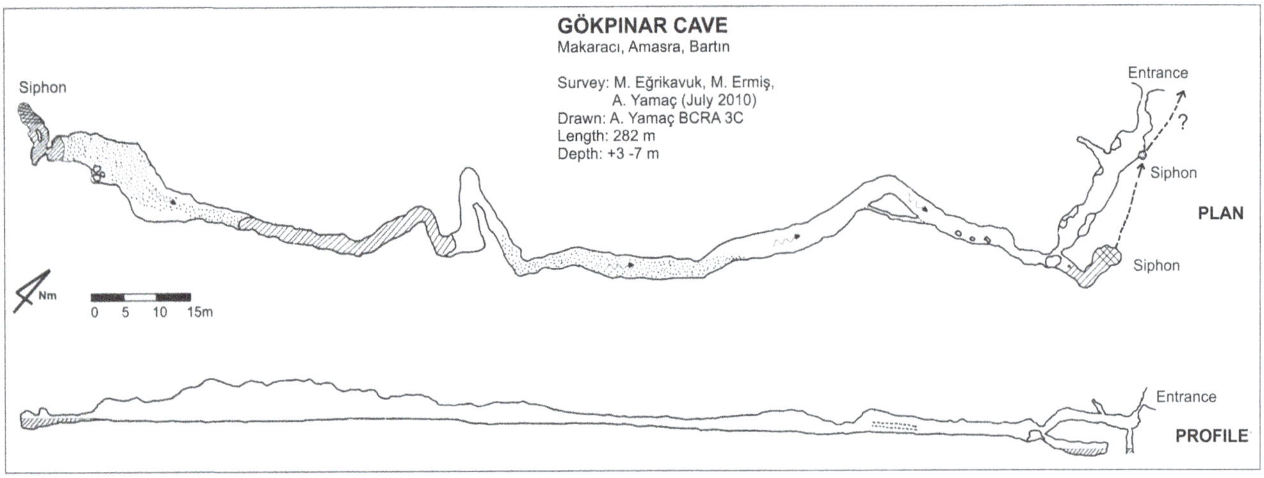

Fig. 3.39 Gökpınar Cave plan and extended profile (drawn A. Yamaç)

3.3.18 Gölcük Cave (No.46)

This cave is located on the east side of Alıçlı Köyü, 500 m away from the road, and is covered with vegetation. The

Fig. 3.40 Gökpınar Cave, cave formations (photo A. Yamaç)

entrance is a 70-cm-diameter hole in a dense forest, which is quite difficult to find under normal conditions. Gölcük Cave, which is reached through this hole that descends about 6 m down at a slope of about 60°, consists of a large, single chamber that is about 28 m in diameter, with the ground covered in clay and mud. Two paths that are located on the west and east sides of the chamber are inaccessible.

Although it is not very rich in terms of formations, small stalactites and stalagmites can be seen around. This is a complete fossil cave whose development stopped a long time ago (Figs. 3.41 and 3.42).

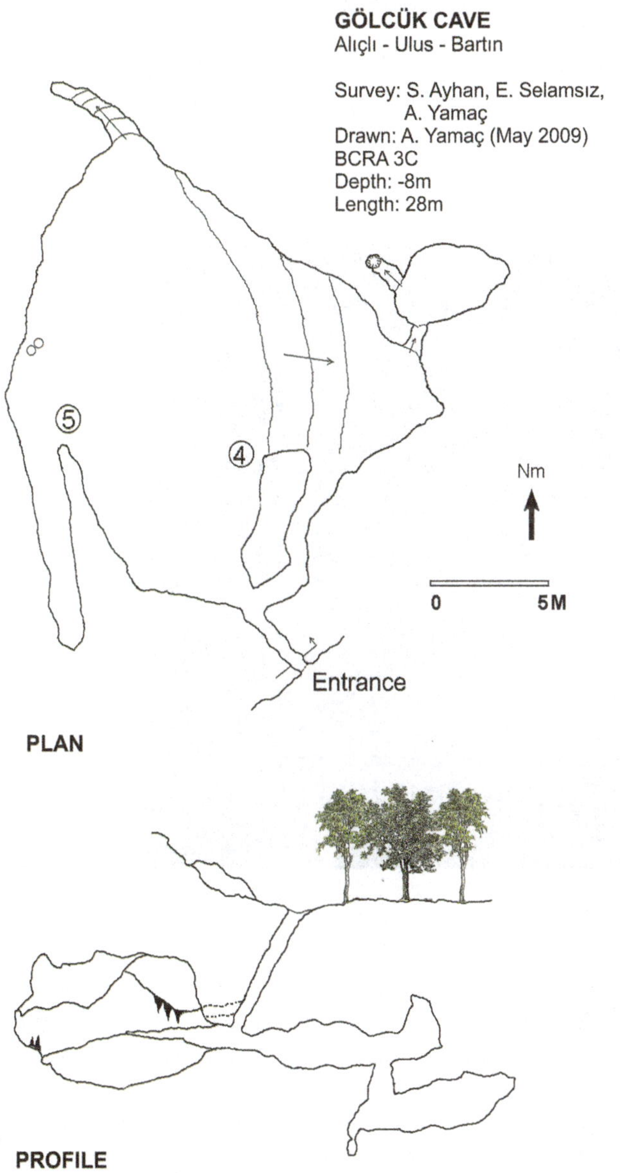

Fig. 3.41 Gölcük Cave plan and extended profile (drawn A. Yamaç)

3.3.19 Gurbettepe Sinkhole (No.47)

The Gurbettepe Sinkhole, situated between the villages of Dizdarlı and Zelle, represents one of the most impressive caves in the region. The cave is situated to the southwest of Sorkun Sinkhole and collects water from a 2.38 km^2 closed basin. The two streams, which are located in the center of the basin and which give the cave its name, merge in the south of the hill and subsequently sink into the cave. During the summer months, the western stream completely dries up, while the eastern stream exhibits a markedly reduced flow. The entrance of the cave is notable for its dimensions (Fig. 3.43). The entrance, where the stream descends as a waterfall, can be reached by traversing the steep slope on the side. The presence of large tree trunks that have been washed away provides an indication of the stream's flow rate during flood periods. One hundred meters from the entrance, a prominent chimney that descends from the left side contributes to the impressive character of the cave. It is most probable that this fossil chimney represents the former entrance to the cave, where the stream once flowed.

In the vicinity of the upper mouth of the chimney, there are a number of small fossil caves. As one progresses through the cave, the light that enters through the imposing entrance and the chimney gradually diminishes. At the end of the hall, a narrow tunnel can be accessed by bending. The 1–1.5 m thick layer of clay visible on the right before entering the tunnel indicates that the tunnel is not sufficient to drain the water during rainy periods. During such periods, the entrance hall of the cave may become a lake, which explains how the giant logs were able to move towards the tunnel. Upon traversing the tunnel, which is inundated to a depth of approximately knee-level, the initial technical descent is reached. Another branch emerges from the entrance hall, leading to the "Pebble Room" at the bottom of the descent. Although the chamber is designated as the "Pebble Room," it is uncertain whether the term "pebble" is an accurate description of the round stones, some of which are 40–50 cm in diameter. It is possible that these stones, which are not found elsewhere in the cave, were shaped by the flowing water as they hit each other. In addition, there are many small pools and formations in the cave as it progresses.

The 710-m-long cave ends with a 7-m-long siphon at a depth of −243 m. However, there are two unexplored branches at −142 and –225 m. It was explored by BUMAK in 1988 and published in Delta 4 (Albukrek, 1993b) (Fig. 3.44).

3.3.20 Gürcüoluk Cave (No. 48)

The Gürcüoluk Cave, which has been designated for tourist access, is situated on the periphery of a ridge that separates

Fig. 3.42 Gölcük Cave, the main chamber (photo A.E. Keskin)

Fig. 3.43 The entrance of Gurbettepe Sinkhole (photo M. Albukrek)

two expansive dolines, approximately 2 km east of the village of Makaracı in Amasra and 32 km from Bartın.

The Gürcüoluk Cave and its surrounding area were formed within Upper Jurassic-Lower Cretaceous-aged limestones that overlay the Permo-Triassic impermeable units. The uvalas in the immediate vicinity of the cave are aligned in the same direction, with the alignment being dependent on the presence of a fault with an NE-SW trending orientation. The area in the vicinity of Gürcüoluk Cave is characterized by a variety of geological features belonging to the Miocene, Pliocene, and Quaternary periods. The highest of these systems, which reach elevations of up to 600 m in short distances, are the erosion surface fragments formed during the Miocene period. The karstification of the area commenced during the Pliocene and continued into the Quaternary. The karstification, which encompasses both paleo- and neo-karst,

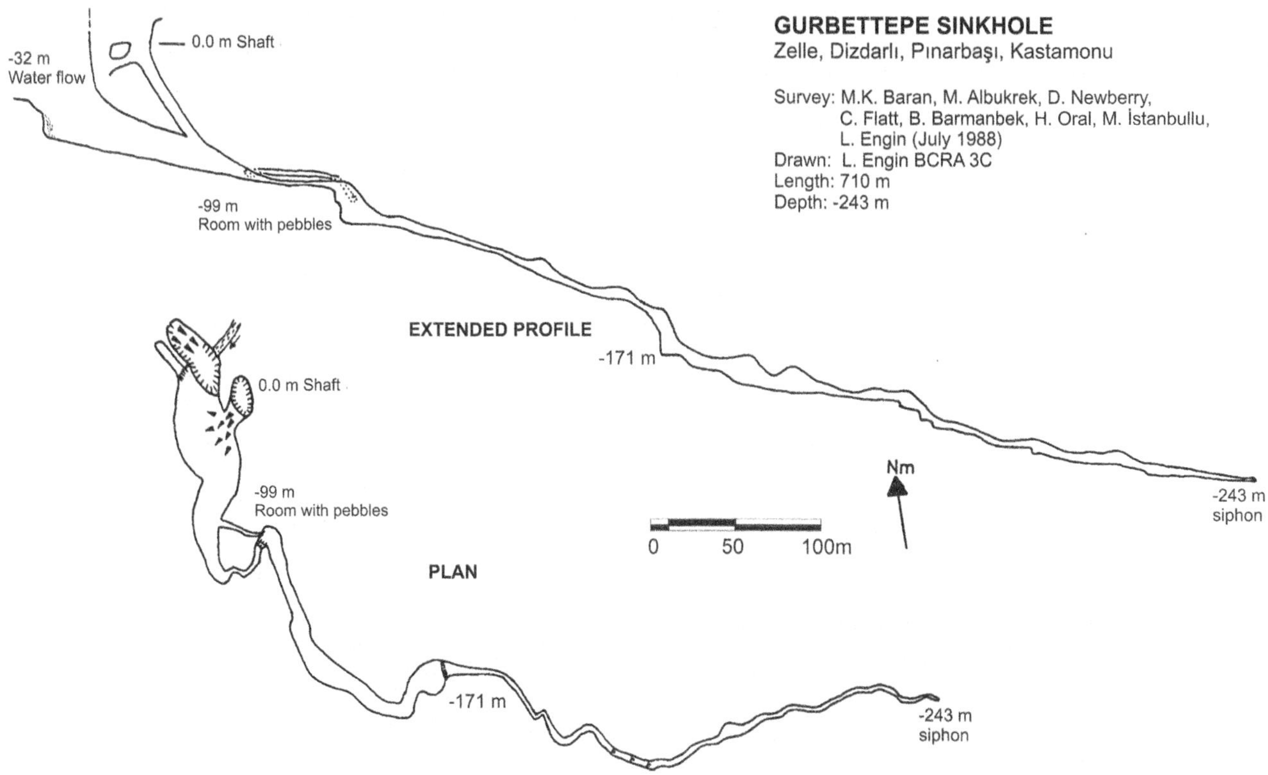

Fig. 3.44 Gurbettepe Sinkhole extended profile and plan (drawn L. Engin)

is relatively shallow. The impermeable rocks situated just below the limestones, which represent the base level, impeded the deeper evolution of karstification. Consequently, poljes, uvalas, dolines, and caves developed in a lateral direction.

Gürcüoluk Cave is a passageway fossil cave comprising a large hall and surrounding chambers, accessed via a roughly triangular main entrance measuring approximately 3 m in width and 2 m in height. Although the cave comprises a single horizontal gallery, the dripstones that developed over time led to the formation of large and small rooms around a hall (Fig. 3.45). Three distinct chambers are situated in the western region of the cave, adjacent to the gallery. At the eastern entrance, a chamber is situated behind the gallery. The total horizontal length of Gürcüoluk Cave is 169 m. The cave has two entrances, one at the eastern end and one at the western end. The eastern entrance is narrower than the western entrance, which is not blocked by soil. Consequently, the western entrance is the primary point of access. The elevation difference between the two entrances is 14 m.

The interior of the cave, which has a width of 2–2.5 m and a ceiling height of 1–7 m, is covered with a variety of formations, including stalactites, stalagmites, columns, wall, and curtain dripstones (Fig. 3.46). These formations, which range in color from gray to cream to beige, contribute to the unique charm of the cave. During the field studies, no specific animal community in the cave was encountered. However, bat communities were observed near the eastern entrance (Nazik et al., 1999). Gürcüoluk Cave, which has completed its development but continues to be shaped inside the cave, maintains its attractiveness due to its beautiful dripstones, easy accessibility, and beautiful natural environment. A detailed scientific survey of the cave has not yet been carried out, and no findings are known.

Since Gürcüoluk Cave was established in 2004 for tourism-related purposes, it has never been properly managed and has been passed from one institution to another. Today, the cave is in an idle state, and many of the structures built for touristic arrangements during this time have been destroyed (Çelikoğlu & Atış, 2015). What is more important is the irreversible destruction of a natural wonder millions of years old as a result of unplanned and visionless decisions made by local or state officials.

Fig. 3.45 Gürcüoluk Cave, the inner chamber (photo L. Nazik)

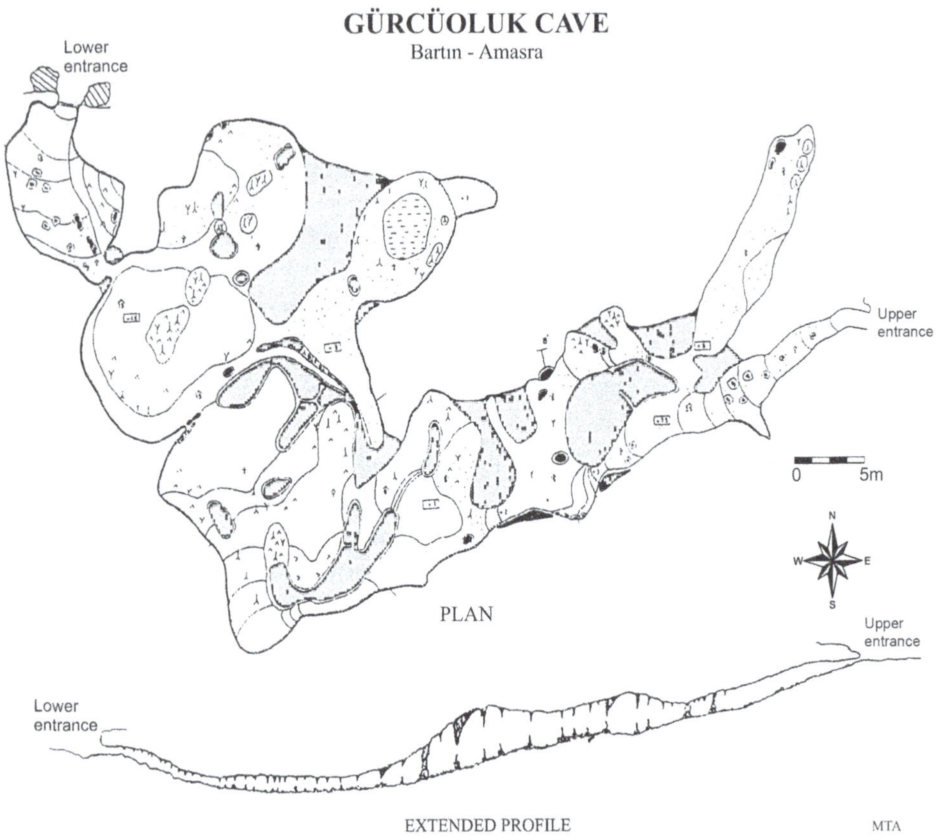

Fig. 3.46 Gürcüoluk Cave plan and extended profile (drawn MTA)

3.3.21 Hamam Cave (No.51)

The Hamam Cave is located 1.5 km to the east of Gideros Bay in Kurucaşile, Bartın. It can only be reached by sea, with an entrance on a vertical wall leading to the sea. The name of the cave derives from the huge freshwater travertine pools inside. The freshwater flow inside the Hamam Cave is hydrogeologically very special, as it is only 2.5 m above the sea and has no entrance towards the land.

Another noteworthy geological phenomenon observed in this cave is its formation in a 20-m-deep fracture in the coastal cliff. The front part of the cave is limestone, while the chamber in the back and the surrounding area are all conglomerates. Despite its relatively small size, this cave is of significant geological and hydrogeological interest and has been designated a first-degree protection status (Figs. 3.47 and 3.48).

3.3.22 Ilgarini Cave (No.55)

The Ilgarini Cave, arguably the most renowned cave in the region, can be accessed following a 4-h hike through the forest from the Topmeydanı District of Sümenler Village. The cave comprises two distinct sections: the fossilized horizontal section and the active vertical section. The horizontal section, with a high ceiling and a total length of 400 m, is divided by a chimney, and the vertical section begins under the chimney. The extensive arm of the horizontal section, which extends to the west after the chimney, is adorned with

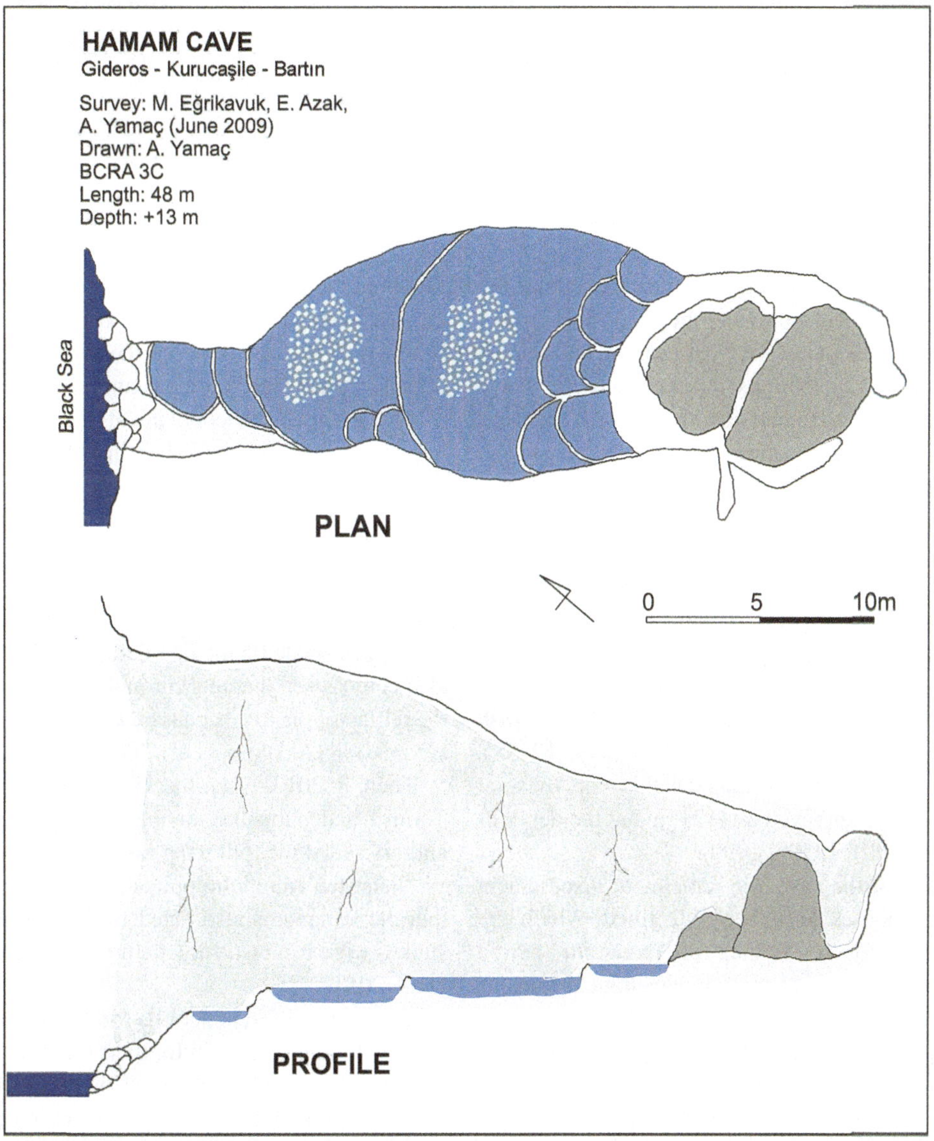

Fig. 3.47 Hamam Cave plan and extended profile (drawn A. Yamaç)

Fig. 3.48 Hamam Cave towards the south (photo A. Yamaç)

stalactites and stalagmites. Three descents necessitate the use of technical equipment following the section where the cemetery remains can be accessed by traversing an inclined branch. The Ilgarini site exhibits minimal activity today. From a depth of −150 m, remarkable travertine formations can be observed in various locations.

Ilgarini Cave was first discovered by Prof. Cemal Arif Alagöz in 1944 and subsequently published on pages 12 and 13 of his work, "Turkey Karst Events" (Alagöz, 1944). The cave was later explored, surveyed, and mapped by BÜMAK in 1982 and published in Delta 1 (Ülkümen, 1983). On this map, the length of the cave was 858 m and the depth was −250 m (Fig. 3.49). For that year, this depth of −250 m made Ilgarini the fourth deepest cave in Turkey. In 1990, the British team, Leicester University Speleological Society (LUSS), visited the cave and drew another map. On this map, the total length of the cave was surveyed as 1131 m and the depth as −240 m (Holland, 1991; LUSS, 1990).

At the entrance of the cave, the remains of a settlement consisting of ten houses have been identified, which are thought to belong to the Roman or Byzantine period (Figs. 3.50 and 3.51). A number of archaeological remains have been discovered in the southern branch, including graves and temples. This branch has been divided into two levels, with the first level containing six three-storey graves and a temple that have been destroyed by people. On the other level, there are five three-story graves and another destroyed temple. The graves in the cave have three floors. The upper floor is wide, and the graves narrow as one progresses from the middle floor to the lower floor. Wooden beams were utilized between the floors, and the body of the deceased was placed on the beams.

The article "Archaeobotanical and Dendroarchaeological Studies in Ilgarini," published in 2004, determined that the woods used in the dendrochronological analyses from graves belonged to the white oak species. A dendrochronological mean was created with eight samples. This chronology was sent to Prof. Peter Ian Kuniholm of Cornell University, who performed the dating. The eight wood samples taken from the graves yielded a median date of 977 AD. However, it is also emphasized that this chronology cannot give the date of the felling of the tree because the outermost rings of the trees are missing (Akkemik et al., 2004).

Then, in 2013, when the same article was translated into Turkish and published in the sixth issue of OBRUK, the authors added the following note to the article:

"Peter Ian Kuniholm pointed out at a conference in 2011 that the specimens also dated to the 5th-6th centuries, and studies were resumed for a definitive conclusion" (Akkemik et al., 2013).

These two different periods for the graves in Ilgarini Cave provide little justification for their presence. This region was undoubtedly a wild jungle in the past, rendering it unsuitable for habitation. When experts first observed the graves and building remains in Ilgarini, their initial assumption was that some clergymen had fled there after Byzantine iconoclasm

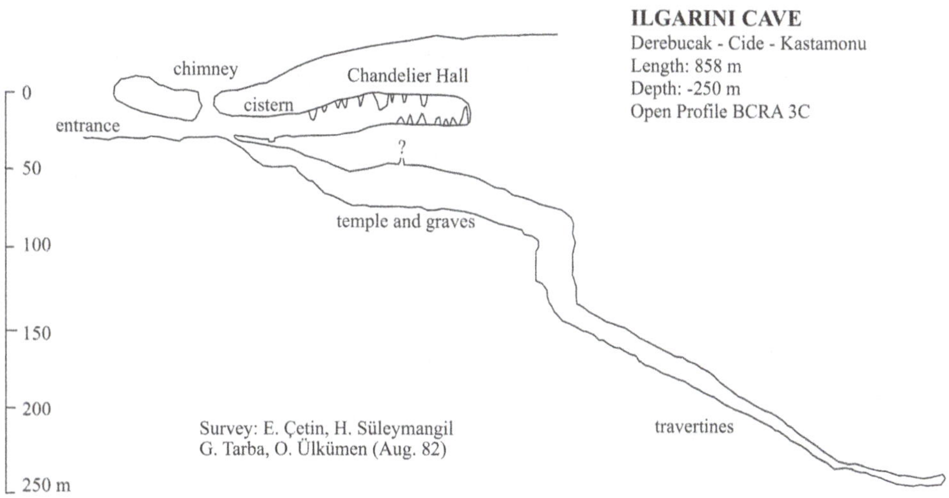

Fig. 3.49 Ilgarini Cave extended profile (drawn O. Ülkümen)

Fig. 3.50 Ilgarini Cave, main chamber (photo D. Lee)

Fig. 3.51 Ilgarini Cave, remains of a medieval structure (photo G. Ambrus)

banned the use of religious images. However, the Second Iconoclasm took place between 814 and 842, whereas the First Iconoclasm, as it is sometimes called, took place between roughly 726 and 787. Therefore, regardless of which dendrochronological date is correct, it can be concluded that the graves found in Ilgarini Cave do not belong to clergymen fleeing from iconoclasm.

A diverse array of biological lifeforms was discovered at the entrance of Ilgarini Cave during the 2009 research expedition. This particular area of the cave is home to several endemic species belonging to the Pseudoscorpionida and Amphipoda groups (ASPEG, 2009).

3.3.23 Inderesi Cave (No.56)

The cave is situated 2 km south of Makaracı Village, Amasra. Despite the region being entirely composed of limestone and exhibiting soil on the surface, the waters escape underground through extensive openings. The initial descent of Inderesi Cave, measuring 5 m, opens into a vast hall. Five distinct entrances are observed on the high ceiling of this 26-m-long hall, with the exception of the primary descending entrance. Subsequently, the waterbed traverses this hall, reaching the second 3.5-m descent through a relatively narrow crack. Following this second descent, the sinkhole narrows after a brief passage and terminates with a siphon (Figs. 3.52 and 3.53).

The cave, like several other caves in this area, is an active or semi-active sinkhole for the majority of the year. The quantity of water extracted during the active period can be gauged from the substantial wood fragments present within the chambers. As no dye test has been conducted in the region thus far, the point at which the sinking waters from Inderesi Cave resurface remains uncertain.

Among the diverse fauna observed within Inderesi Cave, a Gammarus species was identified as a previously unrecorded new species and was named after our club, *Gammarus obruki* (Özbek, 2012).

3.3.24 Intürbesi Cave (No. 60)

The cave is situated on the mountainside in Pınarbaşı, Veray, approximately 1 h's walk from the village. The cave is of considerable size, with a horizontal orientation, and exhibits

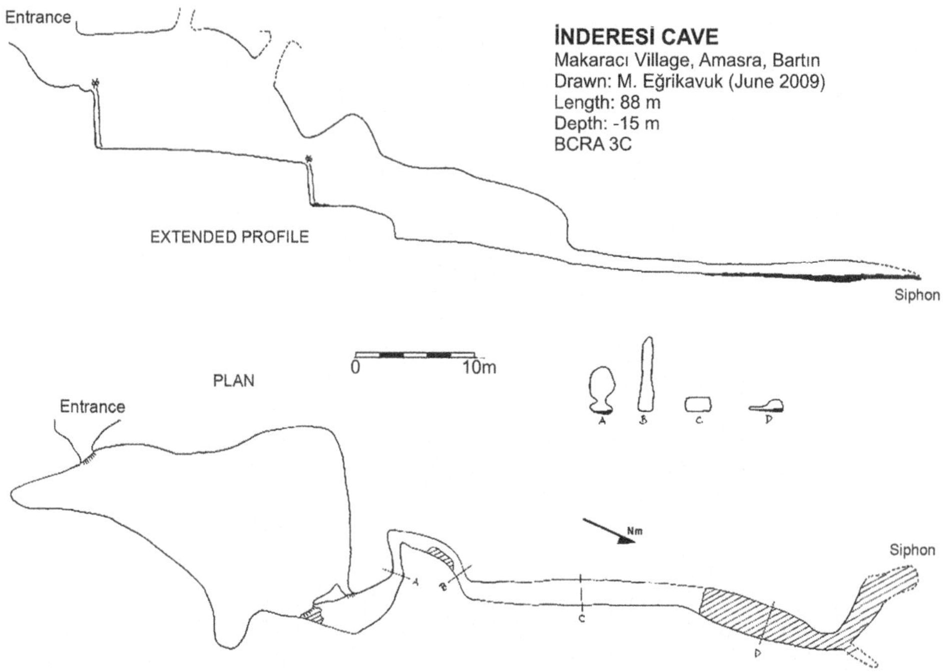

Fig. 3.52 Inderesi Cave plan and extended profile (drawn M. Eğrikavuk)

impressive geological formations. The cave contains numerous indications of past treasure-seeking activities. The majority of İntürbesi Cave is comprised of a single extensive gallery, which attains a height of over 10 m in certain areas and slopes downward. The gallery is notable for its travertine formations and gigantic columns. A descent of approximately 8 m necessitates the use of technical equipment, which leads to a second hall that appears to have been formed as a result of collapse. The flat galleries at the end of both halls narrow and end in impassable locations. Although a gallery was observed on the right side of the main gallery, which might be a continuation of it on the upper level, it was not possible to ascend it.

Intürbesi Cave is a popular tourist destination visited by foreign groups each year. Its name, Intürbesi, which translates to "Cave Tomb" in Turkish, suggests the presence of religious features and historical artifacts dating back to ancient times. Consequently, the cave contains numerous indications of the work of treasure hunters (Figs. 3.54 and 3.55).

3.3.25 Kadıharmanı Sinkhole (No. 61)

The cave is situated on the plateau of Düzköy in Ulus. The Kadıharmanı Sinkhole, which collects water from a very small basin, has an enormous entrance shaft that is disproportionate to the amount of water it collects. The shaft has a diameter of approximately 25 m and a depth of 65 m (Fig. 3.56). The cave continues with a lengthy and uninterrupted gallery, followed by brief descents. After a total length of 428 m and a depth of −132 m, the cave terminates with a siphon (Fig. 3.57). Despite its status as an active sinkhole, the cave exhibits a remarkable biodiversity. Additionally, the suspended vadose zone contains stalactites and stalagmites.

The Kadıharmanı Sinkhole is a predominantly vertical active cave. In addition to the stream that drains the water from a small basin at the main entrance, there are additional water contributions from different points as the cave progresses. These upper branches require further investigation.

Although a significant biodiversity was observed in this cave, regrettably, no biological sampling was conducted, with the exception of a few exceptional specimens. Bats were observed in an inaccessible gallery above the travertine wall.

3.3.26 Kapaklı Sinkhole (No. 62)

The cave is situated approximately 400 m northwest of Topmeydanı, in the vicinity of the village of Sümenler, Pınarbaşı. With a length of 394 m and a depth of −204 m, Kapaklı Sinkhole is one of the 18 active sinkholes that align on the collision point of the limestone mass and impermeable layer here (Fig. 3.34) and has one of the highest flow rates.

Prior to reaching the cave, the Kapaklı Stream infiltrates the ground. It is likely that this water reaches the cave

Fig. 3.53 The first descent of Inderesi Cave (photo A. Yamaç)

through the "Travertine Branch" at the end of the first 24-m descent. During the rainy season, the quantity of water entering the cave is so considerable that the soil sinkhole is unable to accommodate it, resulting in the stream flowing into the cave as a waterfall from the main shaft.

Subsequent to the initial descent, the cave continues with brief descents and minor lakes. The main gallery is consistently wide and high. The presence of tree branches embedded in the fissures at a height of 5–6 m above the cave floor in certain sections of this expansive gallery provides tangible evidence of the considerable volume of water that Kapaklı is capable of draining during the rainy season. The descent commences at a depth of −97 m and is divided into several stations, continuing almost until the siphon at the end of the cave. The initial 59 m of this descent comprise a gallery with a slope of 70° and a completely mud-covered surface (Figs. 3.58 and 3.59).

The cave was explored by BUMAK in 1985–86, and the results were published in Delta 3. During the preliminary exploration of BUMAK in 1985, two species were identified in the lake at −100 m of the cave: *Triturus vittatus* and *Hyla arborea*. In 2009, ASPEG conducted surveys of the cave, during which numerous small species were discovered. The results of this sampling are reported in OBRUK 1.

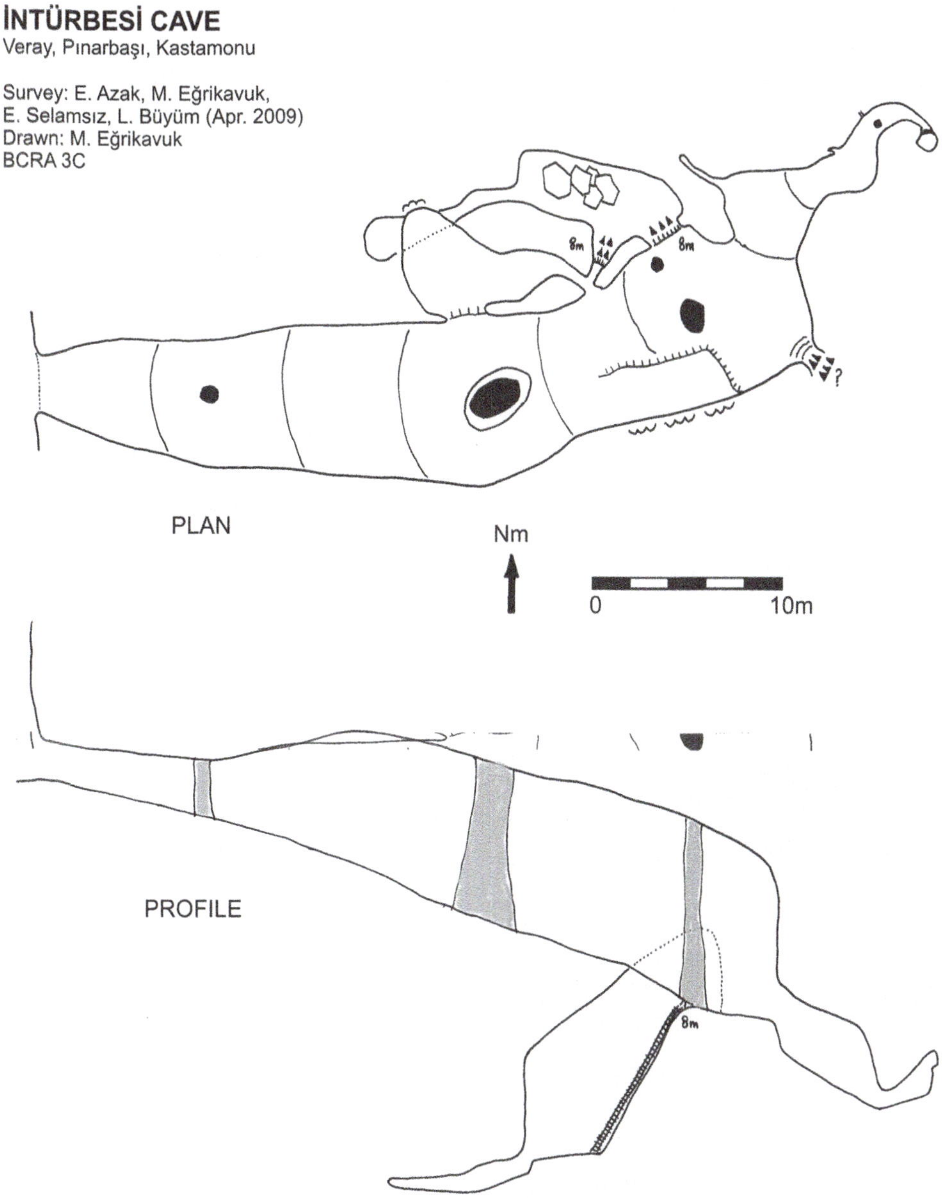

Fig. 3.54 Intürbesi Cave plan and extended profile (drawn M. Eğrikavuk)

3.3.27 Karakaçak Cave (No.63)

The cave is situated approximately 1 h from Karakaçak Village on the Amasra-Çakraz road. It is not possible to locate the cave without the assistance of a guide, as it is situated amidst a dense forest. A gallery with a 20-m slope leads into the cave. The gallery opens into a hall with stalactites. In the hallway, treasure hunters have excavated pits. The cave continues beyond a low, narrow opening on the opposite side of the hallway. The cave ascends at a gentle incline and proceeds through narrow passageways until it reaches the secondary entrance. The cave is 191 m in length and has undergone complete formation. The cave was formed within the karstic Inaltı Formation, which is composed of Malm-Aptian-aged neritic limestones. In accordance with the seasonal precipitation patterns, the water that drips from the ceiling forms small ponds within the cave. In 1993, BUMAK conducted an exploration and mapping of the cave. The report did not mention any findings (Dokay., 2018; Erdem, 1993) (Fig. 3.60).

3.3.28 Karakış Cave (No.64)

This cave, which was reported to be located in the vicinity of Pınarbaşı and whose exact location could not be determined due to deficiencies in the report, was discovered by FACC during their 1992 surveys. The report indicates that the cave

Fig. 3.55 The first chamber of Intürbesi Cave (photo A. E. Keskin)

Fig. 3.56 Kadıharmanı Sinkhole, from the bottom of the first shaft (photo A.E. Keskin)

was discovered after a lengthy trek through a dense, temperate oak/beech forest. The descent from the entrance to the cave floor is approximately 26 m. The entrance ladder descends to the floor of an immense chamber, measuring 40 m in width and 100 m in length. Some of the cave's most attractive features were adorned with white calcite columns and flows. The ice, which appeared to be present throughout the cave, indicated that the temperature was consistently below freezing. No other apparent routes leading out of the unclimbed cave were observed at the point where the cave terminated (Harris, 1992) (Fig. 3.61).

3.3.29 Kaval Sinkhole (No.66)

The cave is situated in Pınarbaşı, Ilıca, in close proximity to Boğazkaya. It is challenging to locate the cave without the assistance of a guide in a region that is densely forested. The cave, which is unidentified in the surrounding villages, with the exception of one or two individuals, and lacks a name, was designated "Kaval Sinkhole" due to its proximity to Kaval Village.

Kaval Sinkhole is an ancient sinkhole that has been preserved in a fossilized state, with a total depth of 54 m and a length of 150 m in two separate branches (Figs. 3.62 and 3.63). The cave is formed by a vertical well, the result of dissolution processes in a crack. Although it lacks an active water carrier feature, the cave is adorned with abundant formations, with the waters leaking from the surface periodically. The cave is notable for its extensive stalactite, stalagmite, column, and especially curtain formations. The

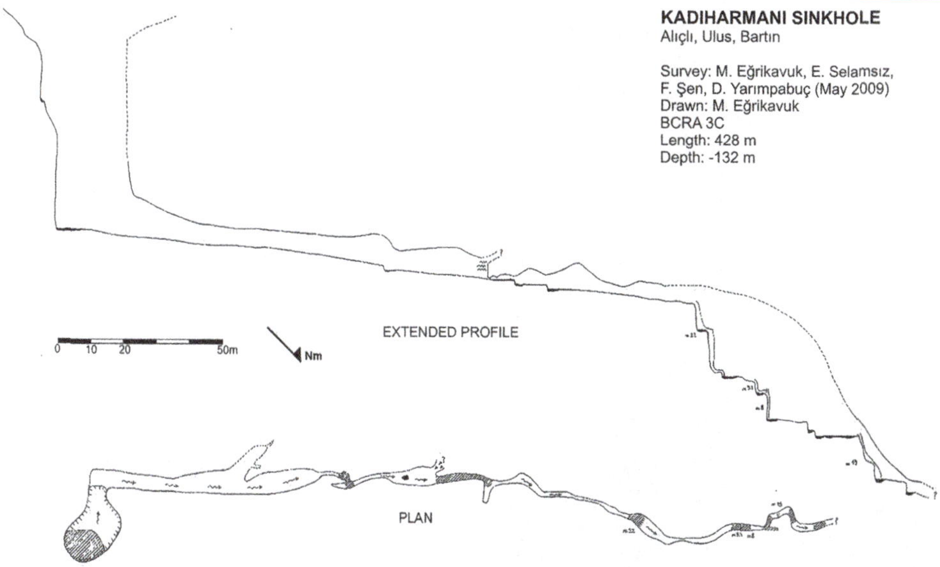

Fig. 3.57 Kadıharmanı Sinkhole plan and extended profile (drawn M. Eğrikavuk)

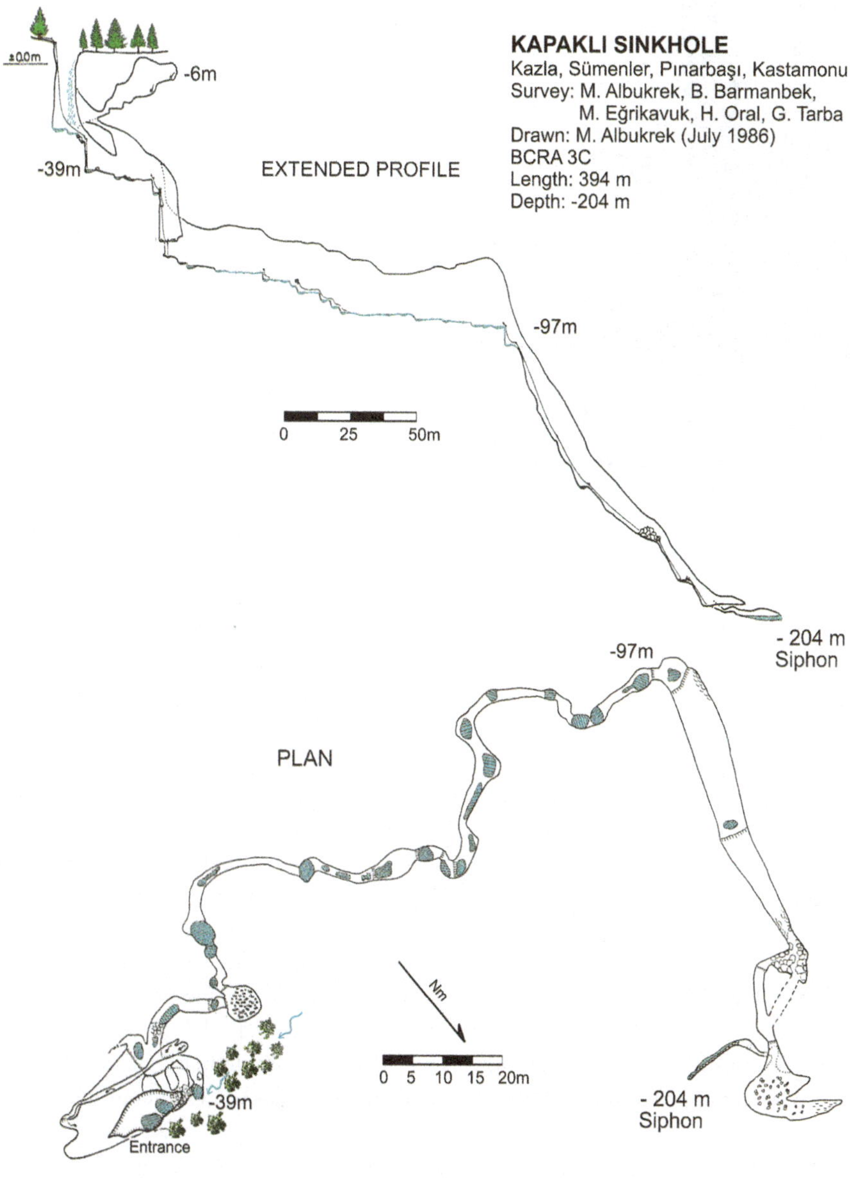

Fig. 3.58 Kapaklı Sinkhole plan and extended profile (drawn M. Albukrek)

Fig. 3.59 Entrance of Kapaklı Sinkhole (photo M. Eğrikavuk)

passages between the vertical descents are also highly inclined and shaped like rock flow with drifting stones.

A few bats were observed throughout the cave. A more intriguing observation was the presence of a particularly active and endearing mouse at the lowest points of the second branch, situated approximately 25 m from the entrance. Nevertheless, despite exhaustive searches, no biological life was observed, with the exception of spiders and flies in the twilight zone near the entrance.

3.3.30 Kayadibi Sinkhole (No. 68)

The cave is located at Pınarbaşı, Sümenler Village, near Valla Canyon in a dense forest cover and appears to be an active sinkhole. From the entrance, there is a steep descent that continues for approximately 120 m. The cave is divided by a natural bridge at the level of 90 m. Although this is an active sinkhole, there are flowstones on the walls in this part. After the main shaft, the cave continues with a slope. In this section, the gallery widens, and the ceiling reaches a height of 7–8 m in places, while the floor is filled with rocks, some of which are larger than 1 m. At a depth of −170 m, a small amount of water enters through a hole at the junction of the floor and the wall. This hole, which is between the piles of stones and cannot be entered, shows that the stream that sank here found a new way to itself. Furthermore, the absence of a sump at the end of the cave provides additional evidence. The sinkhole terminates with debris rocks at a depth of −194 m. The Kayadibi Sinkhole was explored by MAD in 2007 and published in their bulletin No. 14. Despite its depth and importance, the technical information provided in the report is insufficient (Yıldırım, 2008) (Fig. 3.64).

3.3.31 Kılıçlı Cave (No. 72)

Located near Evliyaharman Rock of Çamdibi Village of Cide District, the cave was developed within Upper Jurassic—Lower Cretaceous aged limestones. The two stepped halls at the entrance have travertine and an average ceiling height of 8 m. Kılıçlı Cave, which is 300 m long and developed completely horizontally, is a semi-active cave explored by MTA in 2008 (Törk et al., 2008) (Fig. 3.65).

Kılıçlı Cave has developed in a horizontal direction, with the entrance section collapsing due to fracture systems. It comprises three main halls and galleries connecting these halls. Like many karst caves, Kılıçlı Cave has undergone several stages of formation and reached its present state. During this process, the phreatic zone cave remained completely in the ventilation zone following the lowering of the floor level. With the realization of this situation, the expansion process of the cave ceased, yet the leakage of water from the cracks in the cave indicates that the formation process of the dripstones persists. In the third stage, the blocks torn from the ceiling of the cave descended into the chambers, thereby indicating that the cave has entered its maturity period (Fig. 3.66).

According to the measurements taken in August, while the temperature at the entrance of the cave is 23 °C, it drops to 18 °C in the first hall and 15 °C in the second hall, and it decreases to 10.7 °C in the third hall of the cave. In contrast to the temperature, the humidity in the cave increases as one progresses to the inner parts of the cave. The relative humidity, which was 77% in the first hall, was measured as 80% in the second hall and 84% in the third hall. Kılıçlı Cave is located in the ventilation zone and exhibits a passive cave character, as it lacks a continuous underground flow. Despite

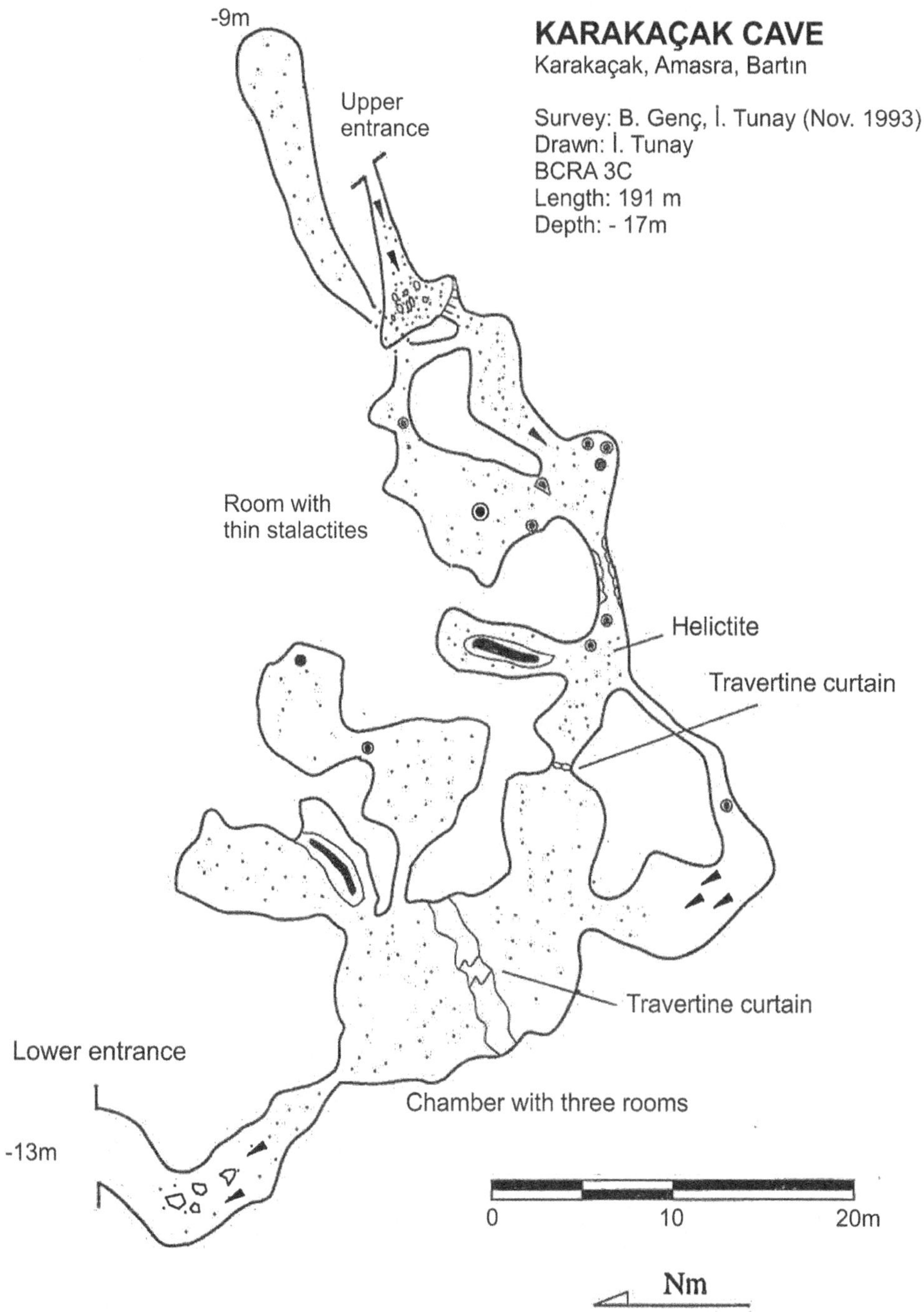

Fig. 3.60 Karakaçak Cave plan (drawn I. Tunay)

this, the cave becomes quite active in the spring months and is therefore unsuitable for habitation. In contrast, a fossil upper gallery is referenced in the Cide Archaeological Project reports (Fig. 3.67). This upper gallery is not depicted on the MTA map. The entrance of the lower gallery is readily accessible from the valley, but the upper gallery of Kılıçlı Cave can only be reached via a challenging ascent, and its geographic position suggests that this gallery was also difficult to reach in the past. This prompts the question of the nature of this site, which is very rich in artifacts.

It seems unlikely that Kılıçlı Cave was a settlement site. Instead, it is more probable that it was a burial or storage cave during the Prehistory and the Bronze Age. In Kılıçlı Cave, some bone fragments were possibly human, and this, in conjunction with isolated finds of a bead and a spindle whorl, could support the idea that this is a disturbed cemetery.

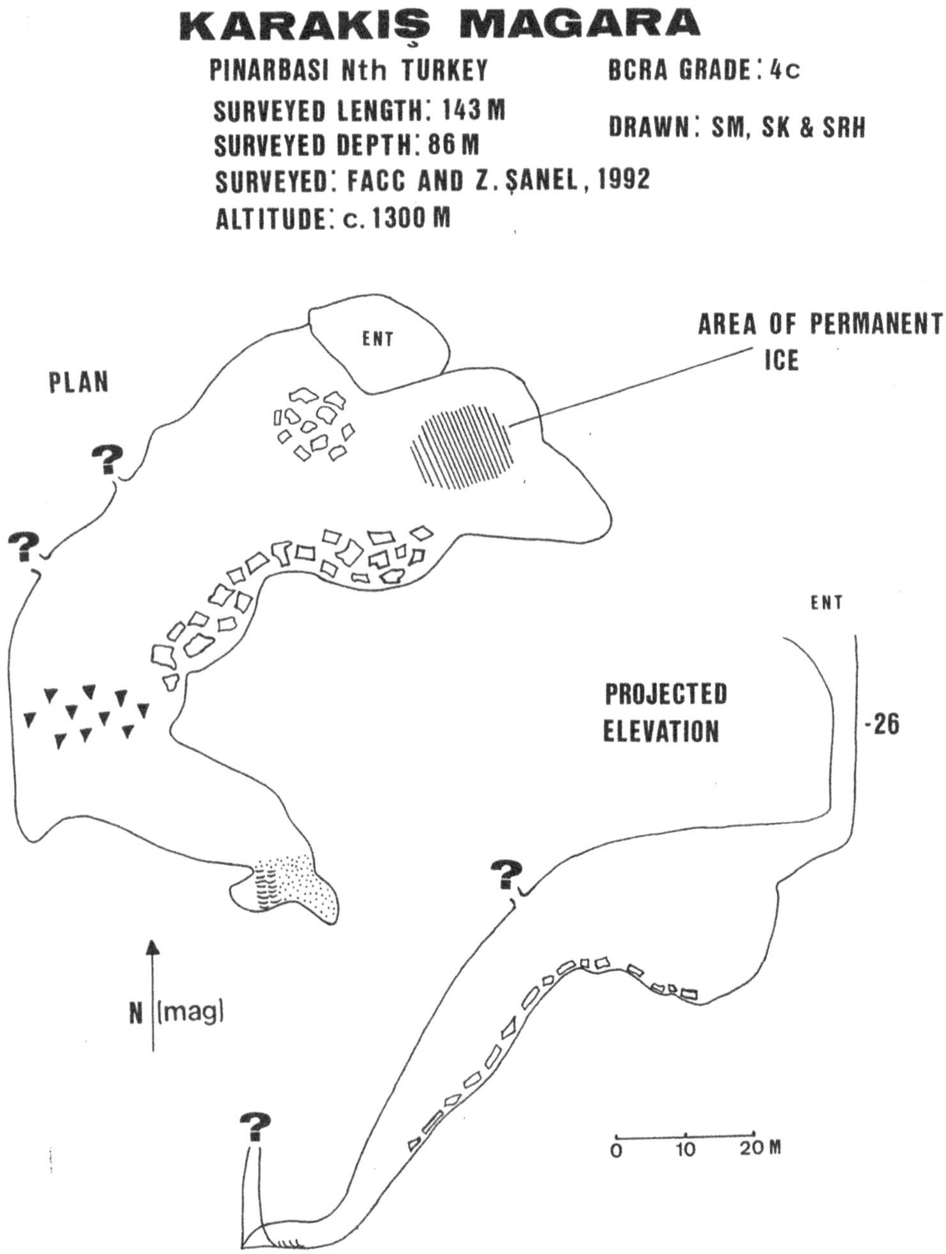

Fig. 3.61 Karakış Cave plan and extended profile (drawn FACC)

The cave exhibited a remarkably dense distribution of materials dating to numerous periods, including the Chalcolithic and Early Bronze Age, which constituted the dominant phase, the second millennium BC. Additionally, some Iron Age and Byzantine period ceramics were discovered.

The richness and diversity of the Chalcolithic and Bronze Age pottery assemblages from Kılıçlı Cave also matched those of Okçular Cave (3.3.39, No. 94). The primary period of use of Kılıçlı Cave is dated to the latter half of the third millennium BC. During this period, the cave yielded a substantial and remarkably diverse assemblage of handmade, well-fired pottery, often decorated with plastic bands and exteriors that were burnished and colored in a variety of hues, including purplish red, pale pink, and yellow (Düring & Glatz, 2015; Glatz et al., 2011; Törk et al., 2008; Törk et al., 2010).

3.3.32 Kocareis Cave (No. 74)

The cave is situated in Makaracı Village in Amasra, Bartın. It is located on a steep slope and has a length of 321 m. It is entered through a water outlet, has very large galleries, and contains numerous living creatures. Some of the village's

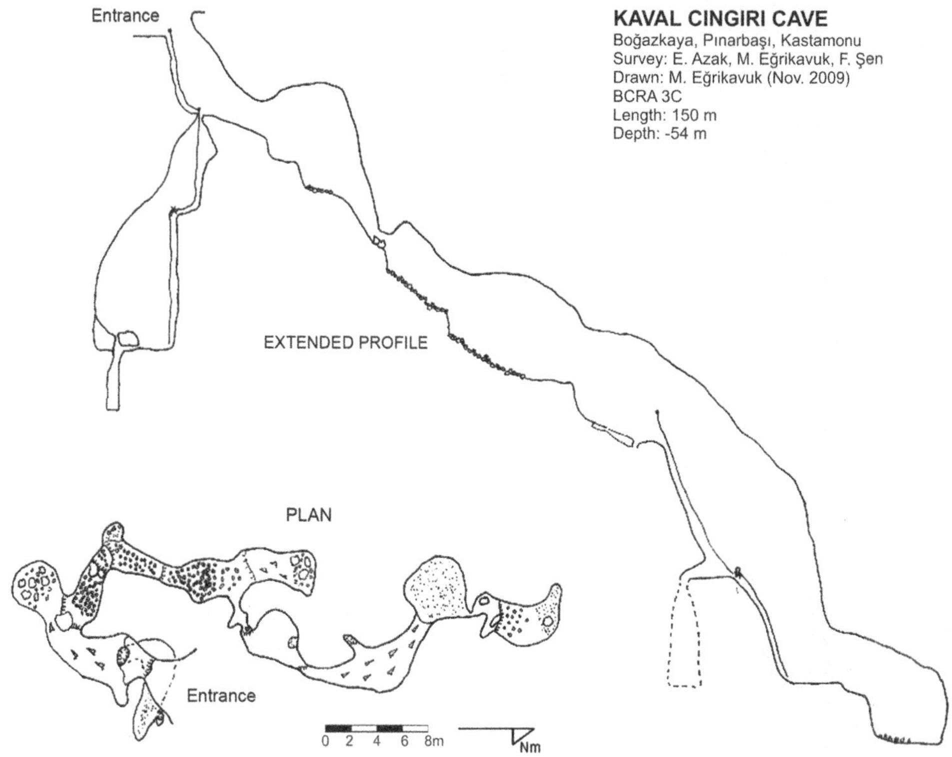

Fig. 3.62 Kaval Sinkhole plan and extended profile (drawn M. Eğrikavuk)

drinking water requirements are met from this cave when it rains a lot due to a powerful water outlet. Despite the dry season during which the exploration was conducted, the cave's narrow passages exhibited significant ponding, with the final 100 m entirely submerged (Figs. 3.68 and 3.69). The stream that emerges from a siphon in the final section is believed to originate from either the earthen sinkhole 400 m to the north or from Gökpınar Cave, which is located behind the hill where the cave is located. In particular, the large parts of the cave are characterized by the presence of numerous flowstones.

Despite its status as an active aquifer, the cave exhibits a high degree of biodiversity, with two adult and one juvenile bat (Myotis myotis) observed in addition to numerous other creatures. However, the presence of guano on the cave floor suggests the possibility of a small colony of bats residing within. While the guano deposits are located in the active water line, they reach a thickness of 1 cm. The cave was first discovered in 2009 and subsequently explored by OMAG in 2010.

3.3.33 Kurtyolu 1 Sinkhole (No. 78)

The cave is situated in the western region of Yaylaköy, within the Armutlu plateau, to the east of Küre Mountains National Park. This is the sole portion of the national park accessible by vehicle and the only area not comprising a dense forest (Fig. 3.70). The cave and two sinkholes were identified and subsequently named "Kurtyolu," a term derived from the region's historic designation, "Kurtyolu." Given that only a small portion of the plateau is accessible and the remainder is a dense jungle, the area is highly risky for exploration. At numerous points, we observed that GPS devices were unable to function due to the complete obstruction of the sky.

Kurtyolu Sinkhole No. 1, which collects water from a very large basin, is a 586-m-long cave formed in covered limestone. A number of the stream entrances are obstructed by surface soil, with only one opening. Despite the lack of water during the season of our exploration, it is evident from the landscape that this opening was insufficient to draw high-flow water during the high season, resulting in ponding. The cave exhibits an almost horizontal formation, extending in a long and narrow gallery. A few side branches along the gallery are obstructed by mud. The cave narrows at the final section and terminates with a sump measuring 7 m in width. The cave was first discovered by ASPEG in 2008 and subsequently measured and mapped in the following year (Fig. 3.71). A few isopods and spiders are known to inhabit the area in proximity to the cave entrance. No life was observed in the following galleries.

Fig. 3.63 The last descent of Kaval Sinkhole (photo M. Eğrikavuk)

3.3.34 Kuşkayası 1 Cave (No. 81)

The cave is situated approximately 2 km east of Göksu District in Yukarıdere Village in Bartın—Ulus. It is located on a hill known as Kuşkayası. Due to the difficulty in locating the entrance, the cave is considered to be quite inaccessible. The cave begins with a small opening, approximately 40 cm in diameter, and a gradual descent. It is notable for its distinctive geological formations and structural features.

Although the cave is a vertical sinkhole formed in a fault fracture, it has different structures at three different levels. It is likely that Kuşkayası 1 Cave was once an active sinkhole, but has since been suspended and fossilized as a result of tectonic movements.

Subsequent to the narrow entrance, a descent of approximately 12 m leads to the initial expansive chamber, which measures between 8 and 10 m in width. The hall is notable for its profusion of formations and the size of its columns. Three distinct descents can be observed from this hall. One of the descents leads to a hall that does not continue. Two large lakes are located in the gallery, which is accessed from the other two via a series of interconnected chambers.

A second gallery can be observed at a depth of 10 m below the aforementioned lakes, which we may designate as the Second Gallery. The total length of Kuşkayası 1 Cave is 155 m, with a depth of 43 m. Despite the absence of evidence of biological activity within the cave, Kuşkayası 1 Cave is notable for its extensive stalactite and stalagmite formations, as well as its diverse cave formations. It is challenging to find a wall without travertine or a floor without stalagmites.

In order to avoid damaging the cave, which is adorned with intricate formations, we exercised great caution while traversing the area and descending. Only one bolt was driven during the 55-m descent (Figs. 3.72 and 3.73).

3.3.35 Kuşkayası 2 Cave (No. 82)

The cave is located in the limestone cliff known as Kuşkayası, approximately 2 km east of Göksu District of Yukarıdere Village in Bartın—Ulus. It has a much larger entrance than Kuşkayası 1 Cave and is situated approximately 200 m west of the aforementioned cave. Despite the cave's 8-m-high, easily visible entrance from the highway, it took us nearly an hour to locate it in the thick forest during our initial exploration.

The cave has a very large entrance and was formed horizontally. From the entrance, one encounters a spacious gallery and a lofty ceiling. The gallery narrows in a northerly direction and becomes blocked after a certain distance. In certain areas, the ceiling height in the main gallery reaches 6–7 m. The total length of the cave is 76 m, with the western branch narrowing to a depth of −7 m. The cave was explored and mapped by ASPEG in 2009. A few insects and a few bats were observed within the cave (Figs. 3.74 and 3.75).

3.3.36 Mantar Cave (No. 85)

The cave is situated approximately 30-min walk from Pınarbaşı, Sümenler Village, Topmeydanı, on the route towards Ejder and Ilgarini caves. As a consequence of the growing number of tourists, the route through the forest has gradually been transformed into a trail. This area, which was previously an uninhabited forest, has begun to diminish in size. It is curious that despite its longstanding reputation, the Mantar Cave has received relatively little attention.

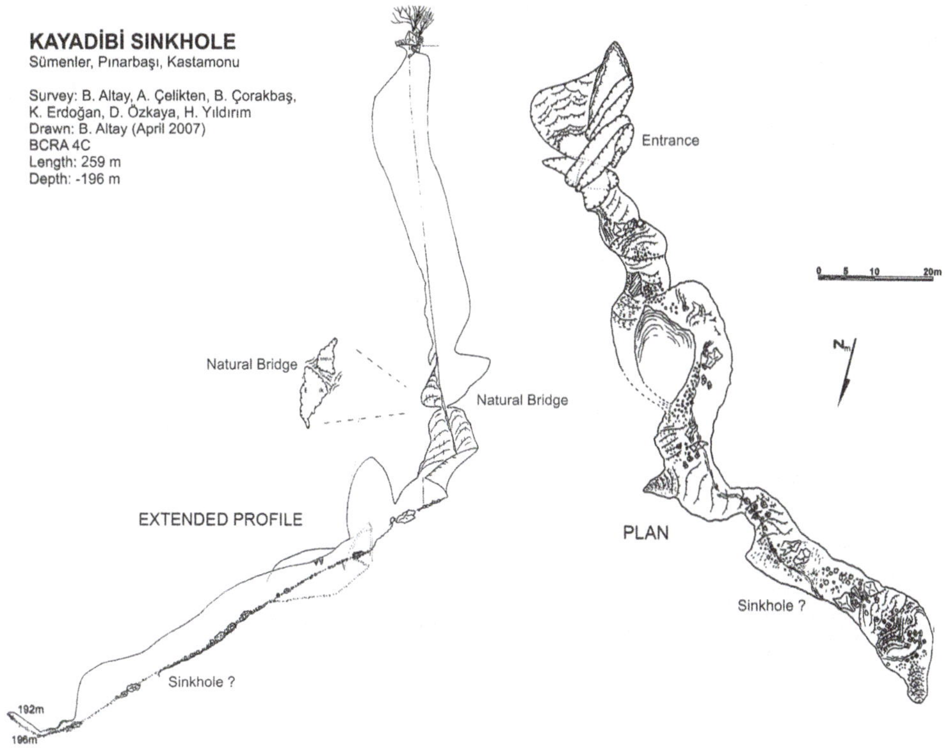

Fig. 3.64 Kayadibi Sinkhole plan and extended profile (drawn B. Altay)

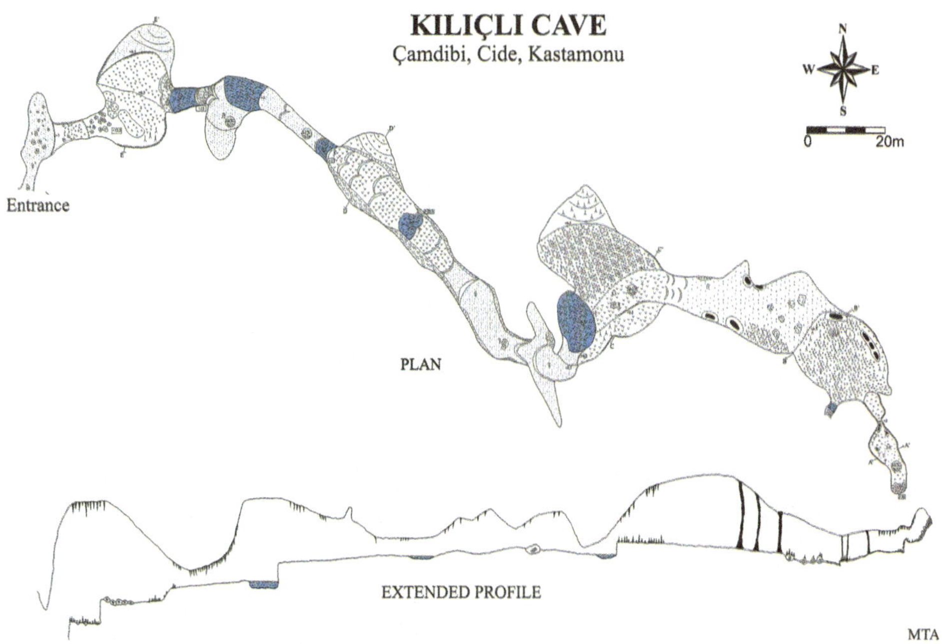

Fig. 3.65 Kılıçlı Cave plan and extended profile (drawn MTA)

Mantar is a fossil cave that has developed on both sides of a depression doline in a densely vegetated Cretaceous limestone. Two large chambers are located to the north and south of the approximately 8-m-deep depression. The total length of the cave is 133 m (Fig. 3.76).

While the northern chamber, which is approximately 40 m long and 20 m wide, lacks any notable formations, the

Fig. 3.66 Kılıçlı Cave (photo E. Kulaksızoğlu)

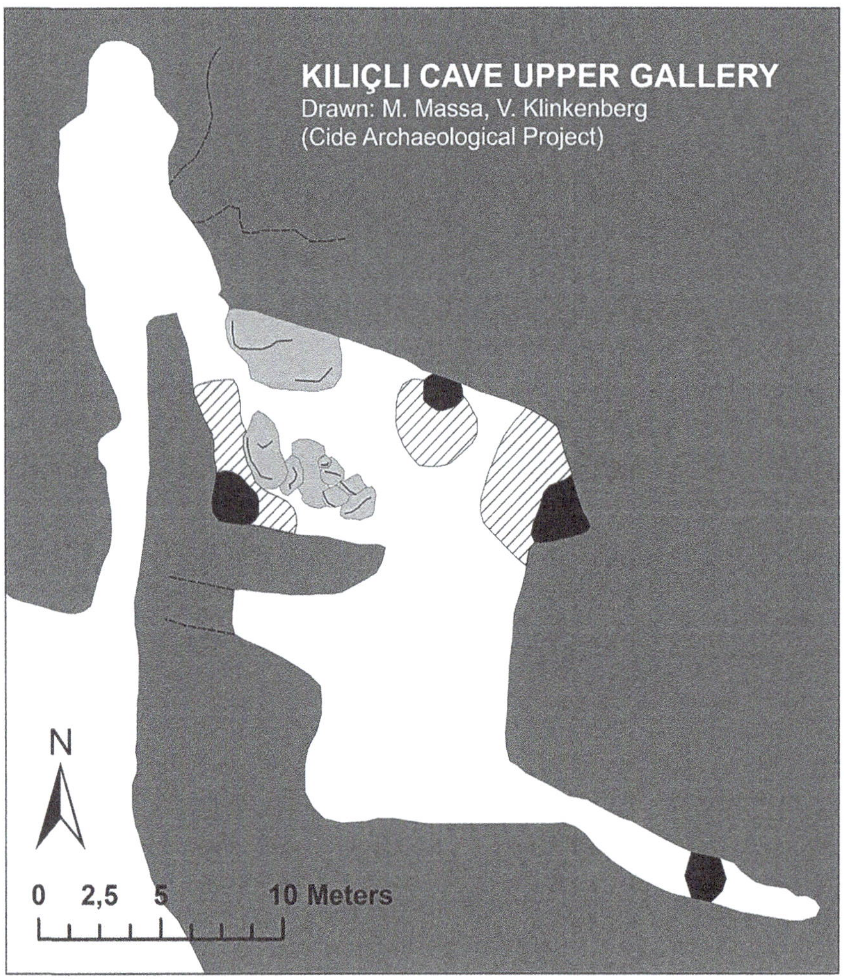

Fig. 3.67 Plan of the upper gallery of Kılıçlı Cave (drawn M. Massa, V. Klinkenberg)

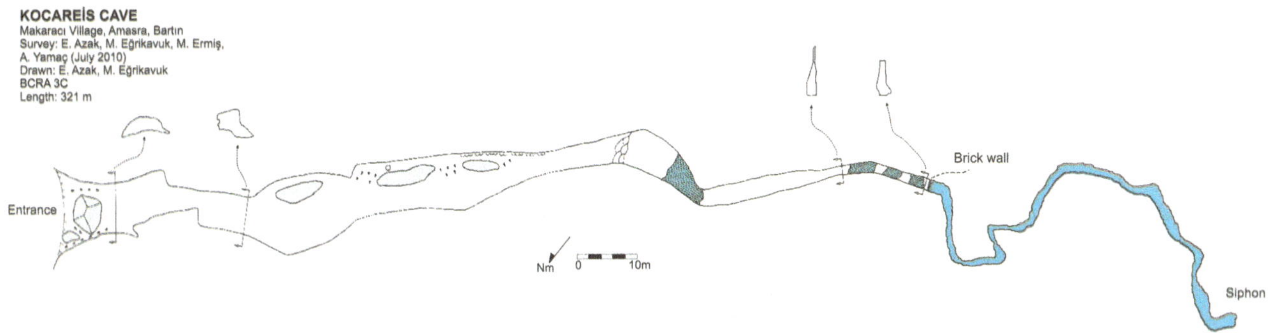

Fig. 3.68 Kocareis Cave plan (drawn E. Azak, M. Eğrikavuk)

Fig. 3.69 Passing one of the submerged passages in Kocareis Cave (photo M. Eğrikavuk)

southern chamber, which is 100 m long and 20 m wide, is adorned with stalagmites and stalactites and has very beautiful formations. Of particular note is a 7-m stalagmite situated in the center of the hall, displaying its full range of layers in the form of a mushroom, which gives the cave its name (Mantar = 'mushroom' in Turkish). In this hall where the ceiling height reaches 15 m at the fourth and fifth stations, some stalagmites exhibit characteristics of columnar structures (Figs. 3.77 and 3.78).

It was explored by ASPEG in 2007 and published in OBRUK 1. In 2009, during the biological studies carried out by ASPEG, a rich biodiversity was found in this cave (Yamaç, 2008).

3.3.37 Medil 1 Cave (No. 86)

The cave is situated in Ahat Village, which is located to the south of Mus District. The cave is accessible via a relatively straightforward 30-min walk from the end of a dirt road leading to the district, which turns left at the end of another road. The cave comprises a vast chamber with a gradual incline, offering no descent or ascent, and a series of short galleries connected to it. The church and cistern are in a state of severe disrepair. The cave has been significantly damaged by treasure hunters. The cave, which measures 205 m in length, features a forest of columns and black stalactites in the low hall at the end of the cave. Additionally, in a small hall near the entrance, piles of roots hanging from the ceiling like stalactites are quite interesting. The cave was initially discovered in 1982 by BUMAK, but it was not thoroughly investigated until 2009 when ASPEG conducted a survey and exploration. With the exception of a few different species and a few bats, no other forms of life were observed in the cave. The remains of a church and a cistern, in addition to the abundant pottery sherds, indicate that Medil 1 Cave, like Ilgarini Cave, was a site of significant human activity in the past (Figs. 3.79 and 3.80).

3.3.38 Medil 2 Cave (No. 87)

The cave is situated approximately 1.5–2-h walk from Örencik Village in the direction of Çatak Canyon. It is not possible to locate the cave without the assistance of a guide, given the extensive vegetative cover in the area.

The cave begins with a depression at the entrance, from which two branches emerge: one wide and the other extremely narrow. These branches lead into a spacious hall.

Fig. 3.70 Armutlu plateau (photo A. Yamaç)

The galleries that cross each other from top to bottom connect the branches. The total length of Medil 2 Cave, which was formed in a fault fracture, is 211 m and is embellished with passageways that ascend and descend. The small, exquisite formations are located in the slender galleries. The cave was initially discovered by BUMAK in 1982, but ASPEG was unable to explore and survey it in 2009. With the exception of a considerable number of spiders belonging to the same species, no other forms of life were observed (Figs. 3.81 and 3.82).

3.3.39 Okçular Cave (No. 94)

The Okçular Cave, one of the most significant discoveries of the Cide Archaeological Project, is situated beneath the Byzantine castle of Okçular Village in Cide, Kastamonu Province. The cave's entrance is particularly narrow. After traversing a narrow passageway, one must descend a precipitous slope of approximately 5 m in height. The cave below is notable for its extensive stalactite and stalagmite formations, as well as its floor, which is covered with rocks from episodes of roof collapse. At the rear of the cave, an area comprising brown crumbly, moist loamy clay was discovered, which was found to contain a considerable number of prehistoric burnished ceramics and lithics. Beyond these accumulations, a stratigraphic profile was identified, comprising horizontally laminated midden deposits containing ash, charcoal, animal bone fragments, flints, and a substantial quantity of pottery.

It appears that a former cave entrance was utilized as a refuse disposal site in the prehistoric period. This would suggest that during the Chalcolithic period, the rock outcrop of Okçular was inhabited by a settled population. Byzantine times saw the outcrop fortified as a stronghold. The rock outcrop, which is largely covered with dense shrubbery, does not exhibit any visible evidence of prehistoric artifacts or features. However, the high density of midden deposits and the considerable quantity of stratified prehistoric pottery in the cave shaft below indicate that the site was occupied for an extended period. It is noteworthy that analogous settlement sites were documented in other regions of Anatolia during the Middle Chalcolithic period.

The discovery of obsidian chipped pieces, albeit in small numbers, is of considerable interest as it provides the opportunity to identify geological sources and trace geographical connections. Based on the available evidence, it can be reasonably concluded that obsidians of Okçular Cave probably derive from the Yağlar source, located near the modern town of Kızılcahamam. Obsidian from the Yağlar source has also been identified in Neolithic sites of the Marmara region, including Ilıpınar, Pendik, and Fikirtepe. In these sites, it co-occurs with obsidian from the Cappadocian sources. Currently, the distribution of Galatian obsidian is

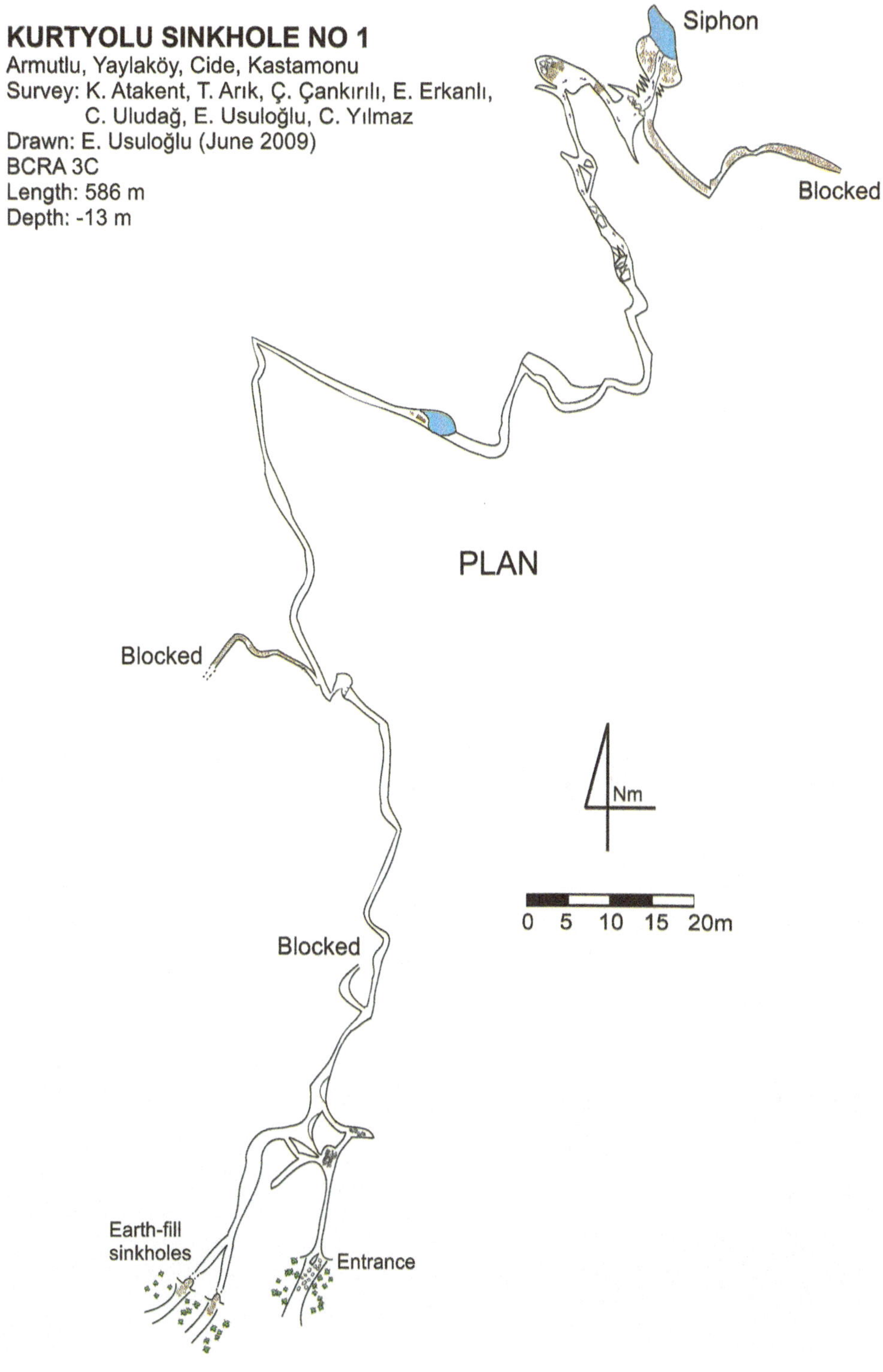

Fig. 3.71 Kurtyolu 1 Sinkhole plan (drawn E. Usuloğlu)

limited to northwestern Anatolia and a brief period in the early Ceramic Neolithic, approximately 6500–5800 BC. It is noteworthy that recent excavations at the Late Chalcolithic Çamlıbel Tarlası, situated in close proximity to the Galatian sources, have yielded obsidian originating from four distinct Cappadocian sources, yet not from the nearby Galatian sources. This indicates that by the Late Chalcolithic, the Galatian sources were no longer at the center of an obsidian exchange network. Conversely, the sample from Okçular Cave, which can be dated with some confidence to the early

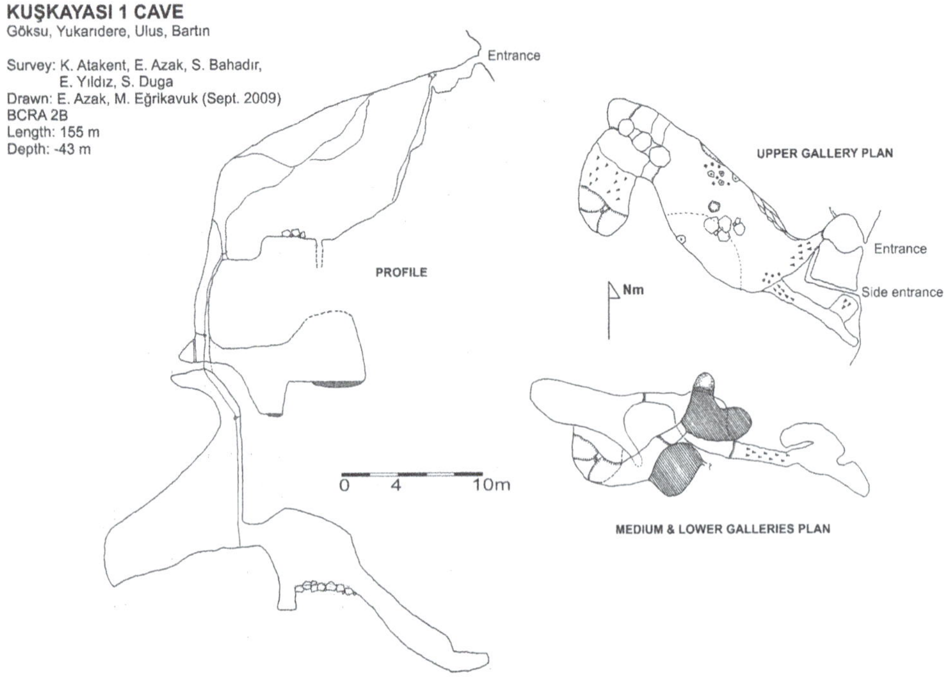

Fig. 3.72 Kuşkayası 1 Cave plan and profile (drawn E. Azak, M. Eğrikavuk)

Fig. 3.73 Kuşkayası 1 Cave (photo A. E. Keskin)

fifth millennium BC, indicates that this obsidian exchange network might have continued to exist at least in part during the Middle Chalcolithic (Düring et al., 2012; Düring & Glatz, 2015; Glatz et al., 2011) (Figs. 3.83 and 3.84).

3.3.40 Önüne Cave (No. 95)

The cave, which measures 220 m in length and 36 m in depth, has not yet been located. It is situated to the east of Cide-Çalköy Village and is believed to be located near Loç Canyon. It was explored by the LUSS team in 1990. The original report is as follows:

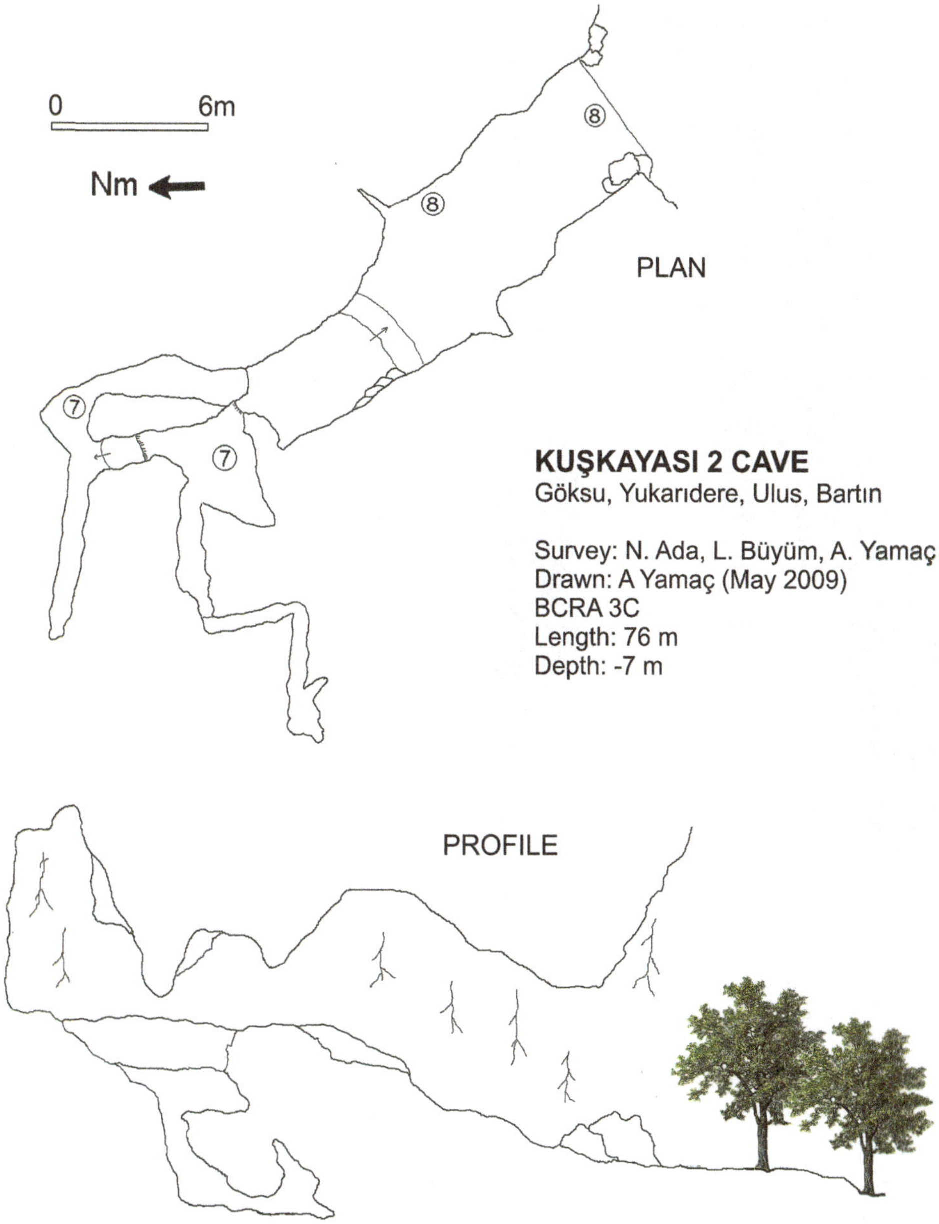

Fig. 3.74 Profile and plan of Kuşkayası 2 Cave (drawn A. Yamaç)

"The entrance of Önüne Mağara (Cave) is about 1.5 km from the village in a due east direction (ask in the village). An open roofed rift continues to lead onto a small chamber containing the remains of some crude stonework shelter.

An 8 m overhang climb can be overcome with the use of some friendly villagers and a newly felled tree to act as a scaling pole. The passage rapidly leads to a similar but easier climb (bolted for the handline on the right). The passage opens into a large chamber with dry gours and many bats (with associated guano). A spacious, well-goured passage leads on for 100 m, where several short cross-rifts enter before a low roofed, 1 m deep pool is reached (can be traversed around on the right). On the other side of the pool, the cave opens out into the superbly decorated hall of the university, complete with deceptive acoustics. At the end of a further, well-decorated passage and second chamber the way is largely blocked by calcite formations but two squeezes rejoin at a further chamber where the way on is finally blocked by a muddy ramp, leading to an overhanging boulder choke. Before the squeezes are reached, on the right of the chamber,

Fig. 3.75 Kuşkayası 2 Cave (photo A. Yamaç)

a wide, low, sandy-bottomed pool leads to an upstream sump of 2 m diameter in clear water. This could be siphoned off into the series of steeply descending gours. However, we didn't have the necessary 30 m of tubing" (Holland, 1991) (Fig. 3.85).

3.3.41 Sipahiler Cave (No.100)

It is located 2 km from Kayadibi Village in Bartın. The cave, which is located beneath dense vegetation and covered limestone, is notable for its unexpected splendor, which is revealed upon entering through its small entrance. The formation of vast speleothems in the Sipahiler Cave, situated in the western region of the national park, can be attributed to the presence of substantial Permian limestone and the associated vegetation.

The chambers, which span three levels, contain numerous gigantic stalagmites. The diameters of the two stalagmites at the bottom of the first hall at the entrance are approximately 5 m each, while the toppled stalagmite at the lowest level is approximately 10 m long (Figs. 3.86 and 3.87). The steeply sloping passage leading down to this lower level and the lowest hall is completely covered with mud.

The cave, which measures 338 m in length, boasts a ceiling height of 20 m in the lower hall. This section is home to not only stalagmites but also a rich variety of formations on the walls.

The cave was explored and mapped by ASPEG in 2009. Sipahiler Cave, which has extremely impressive formations and rich biodiversity with endemic species of the Orthoptera group, is very close to the settlement center and should urgently be taken under protection.

3.3.42 Sorkun Sinkhole (No. 101)

The Sorkun Sinkhole is situated within the Sorkun Plateau of Sümenler Village and serves as a conduit for the waters of this expansive basin. It is the highest-flowing sinkhole in the area and was formed along the intersecting cracks on the southern border of the limestone strip in this basin.

In a series of tests conducted over a period of several days, it was determined that approximately 1200 liters per second of water entered the Sorkun Sinkhole in April. This quantity is greater than the total amount of water entering all other sinkholes in the region during the same period. It is likely that the water source known as the Sorkun Spring, situated on the western wall of Valla Canyon, is the source of the water entering the cave.

The cave, which terminates in a siphon behind a lake, has a total depth of −200 m, as measured thus far. However, there is a possibility that it may continue. In 1983, BUMAK conducted an investigation of the Sorkun Sinkhole, which was subsequently published in Delta 1. The report does not mention any findings in the cave. Conversely, although it is an active sinkhole, a rich diversity of life was discovered during the biological studies conducted by ASPEG in 2007.

In addition, the previously uncharted side branch, as referenced in the earlier report, was subjected to further investigation. As indicated in the report, the aforementioned side branch had never been explored due to the high river flow rate at the time. The side branch was located at the bottom of a 40-m steep descent from the upper entrance and continued as a separate branch from the lower entrance after two descents of 21 m in total. It became evident that this gallery, presumed to be a side branch in the 1983 survey, was in fact larger than a side branch. After a total of 38 m of descents, due to the lack of available material, the survey was terminated in a gallery below this descent, which continued with lakes. The discovery of a second side branch in the middle of the descent and the continued expansion of the gallery at the bottom prompted the question of whether this side branch had turned and connected to the main gallery (Figs. 3.88 and 3.89).

After all, it is certain that the Sorkun Sinkhole continues with several different branches.

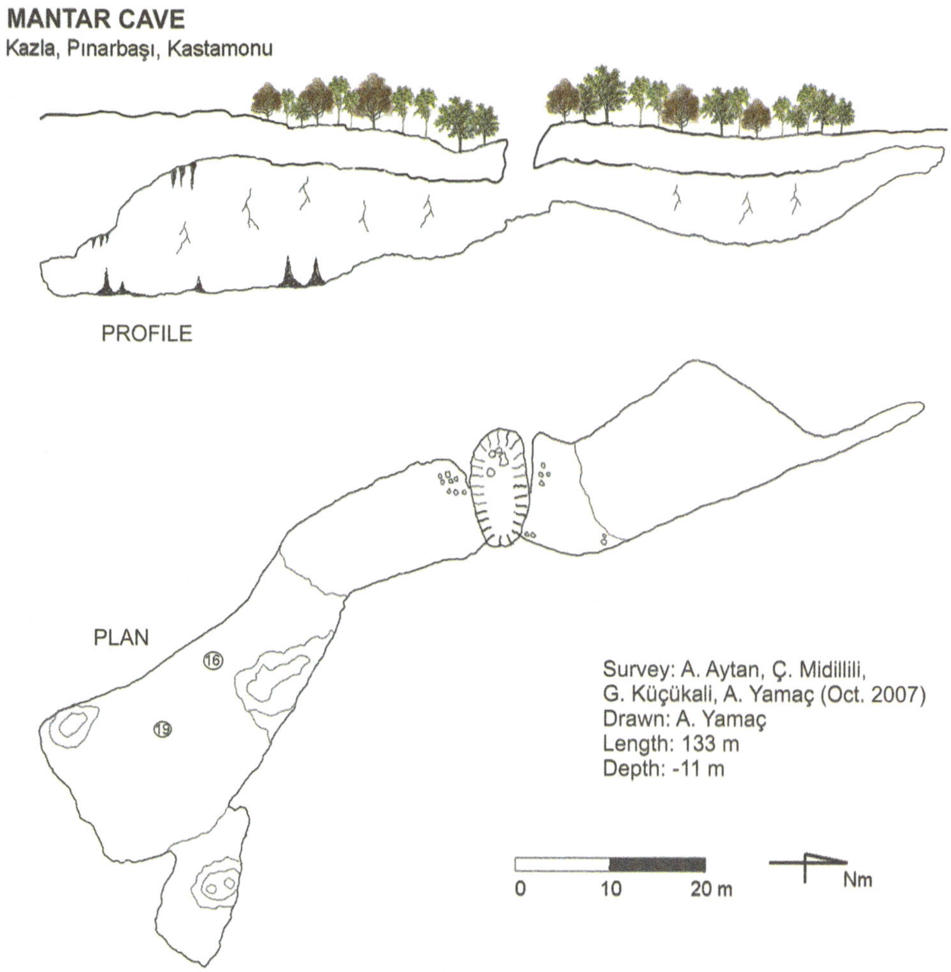

Fig. 3.76 Mantar Cave plan and profile (drawn A. Yamaç)

Fig. 3.77 Mantar Cave (photo G. Ambrus)

Fig. 3.78 Mantar Cave (photo M. Renda)

3.3.43 Şinasi Cave (Gideros 1) (No. 105)

As previously stated in Chap. 2, the inability of the LUSS, FACC, and ADEKS teams to accurately locate the caves they discovered in the 1990s resulted in the loss of some of these caves. This was due to the dense forest cover in the area and the lack of GPS technology at the time. The only cave that does not fit into this understandable situation and that may be considered one of the most problematic in this entire list is the Şinasi Cave. The cave in question is situated on the Gideros Bay of Bartın, Kurucaşile, and is depicted on two different maps under two different names.

In 2009, we conducted an investigation of the cave, situated a mere 20 m from the main road and within a dense forest. We inquired with the local residents regarding the name of the cave and utilized the designation they provided, which was "Şinasi Cave."

Subsequently, the Cide Archaeological Project conducted an investigation, survey, and mapping of the cave in the same area in 2015. The cave was subsequently designated as Gideros Cave 1. In their archaeological report, only a few notes are provided:

"The walls of Gideros Cave 1 are covered in multiple Greek graffiti. Found near the religious complex discussed above, Gideros Cave 1 is a large cave complex. At some point, it was turned into a commemorative shrine, which is noted by an apsidal enclave at the back of the cave surrounded by crosses scratched into the wall. There are also some Greek inscriptions which need further analysis, as they will almost certainly offer some indication of the purpose and dedication of this site." (Düring et al., 2012; Düring & Glatz, 2015).

Although there are differences between the two maps, it is clear that "Şinasi Cave" and "Gideros Cave 1" are the same (Fig. 3.90). There are quite large stalactites and stalagmites

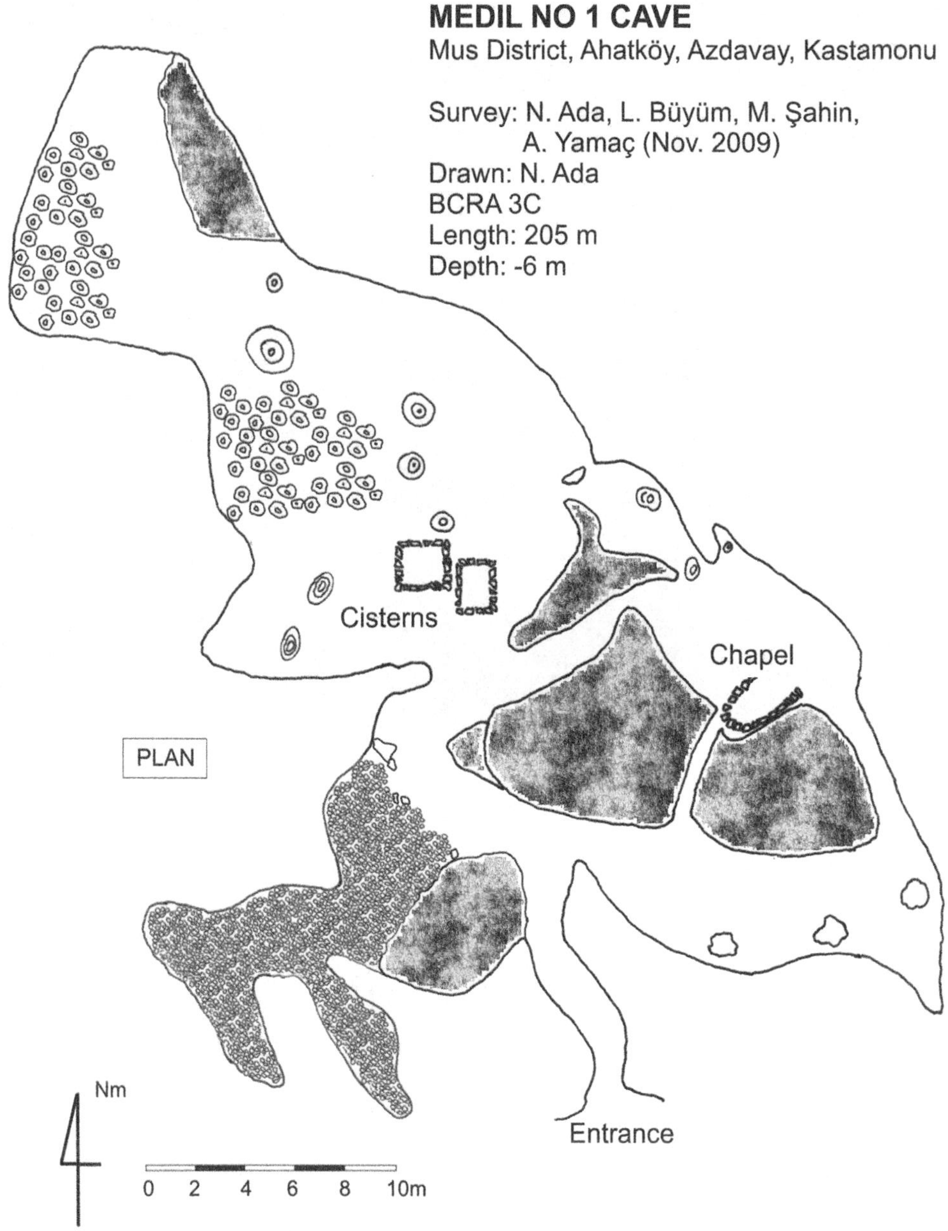

Fig. 3.79 Medil 1 Cave plan (drawn N. Ada)

in the 80-m-long cave, which continue in a straight line as two different chambers (Fig. 3.91). At various locations throughout the cave, treasure hunters dug pits, some of which were quite deep.

3.3.44 Topmeydanı Sinkhole (No. 106)

The Topmeydanı Sinkhole is situated in the Topmeydanı Region, 2 km north of Sümenler Village, Pınarbaşı. The cave, which was previously a sinkhole, has been partially fossilized by being suspended in time. The stream, which previously drained into this area, now exits underground through an earthen sinkhole beyond the cave. Although the cave joins Topmeydanı Kuylucu at a certain point, no active water flow was observed from any point during the surveys.

The majority of the side branches are obstructed by mud, likely due to the fact that a portion of the cave is suspended and therefore unable to effectively absorb water. Conversely, during the rainy season, the drained water accumulates in isolated areas, forming small ponds. The majority of the cave comprises fossil galleries. One of these fossil galleries turns and rises towards the entrance of the cave, where it is covered with very beautiful formations.

Topmeydanı Sinkhole is −77 m deep and the longest cave ever found in the Küre Mountains National Park, with a

Fig. 3.80 Medil 1 Cave (photo M. Şahin)

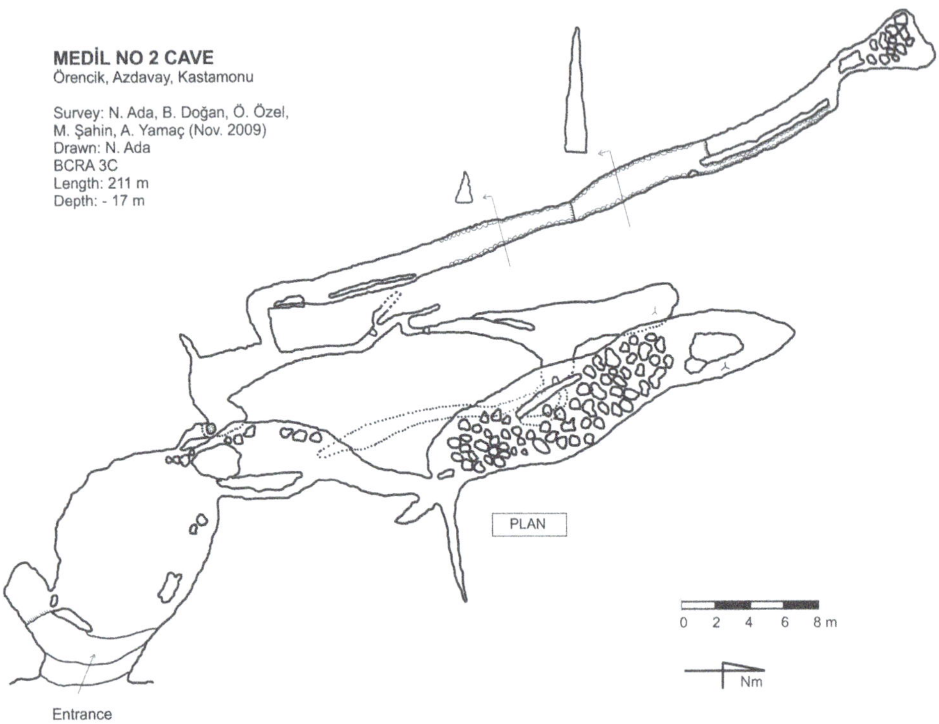

Fig. 3.81 Medil 2 Cave plan (drawn N. Ada)

Fig. 3.82 Medil 2 Cave (photo M. Şahin)

length of 1470 m. The cave was explored by BUMAK between 1984–87 and published in Delta 3 (Figs. 3.92 and 3.93). In 2009, during the biological studies carried out by ASPEG, a very rich biodiversity was found in this cave.

3.3.45 Uçakdibi Cave (No. 118)

The Uçakdibi Cave is situated in the vicinity of Atköprü, Cide, Kastamonu. The cave is located within a dense forest on the border of Küre National Park. The Cide Archaeological

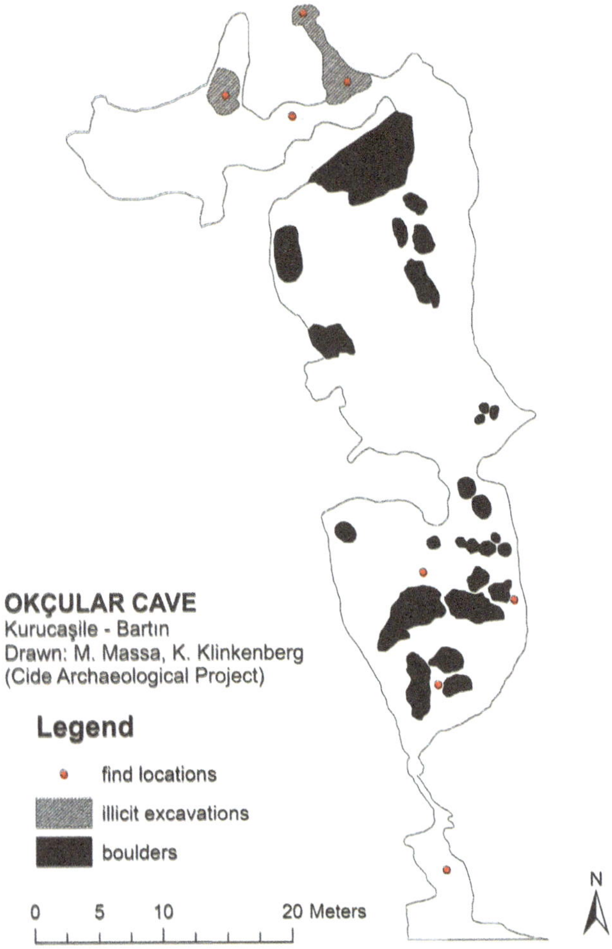

Fig. 3.83 Okçular Cave plan (drawn M. Massa, V. Klinkenberg)

Fig. 3.84 Okçular Cave (photo Cide Archaeological Project)

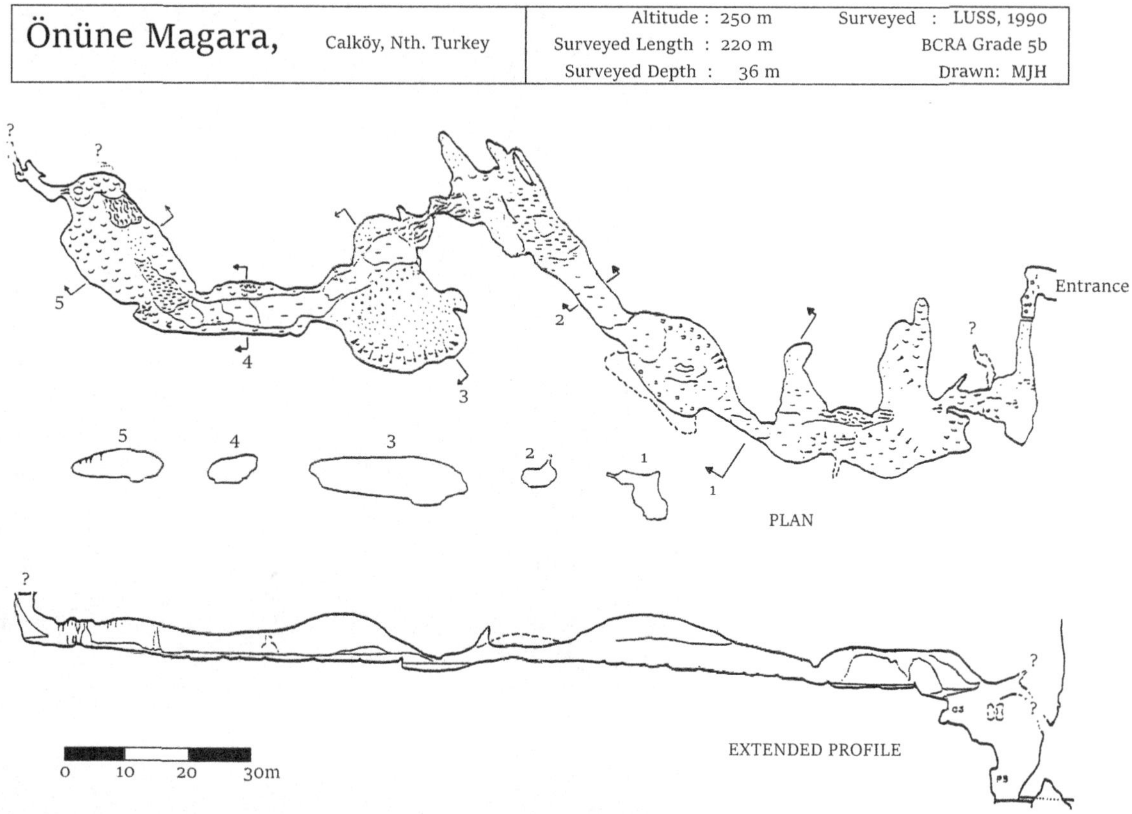

Fig. 3.85 Önüne Cave plan and extended profile (drawn LUSS)

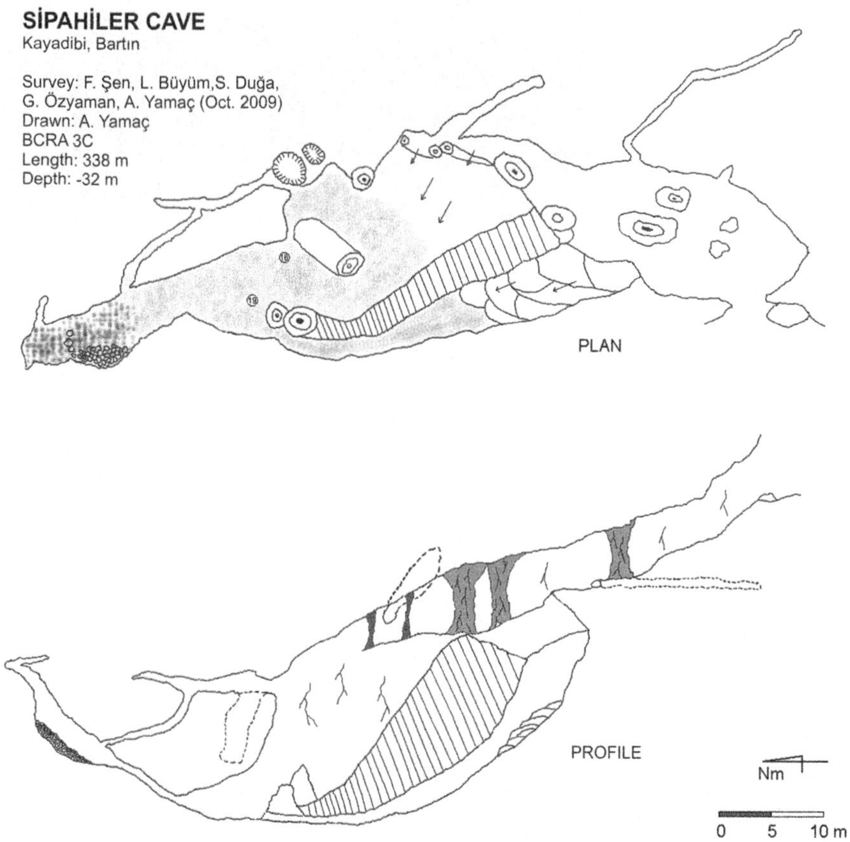

Fig. 3.86 Sipahiler Cave plan and extended profile (drawn A. Yamaç)

Fig. 3.87 Sipahiler Cave (photo Ç. Çankırılı)

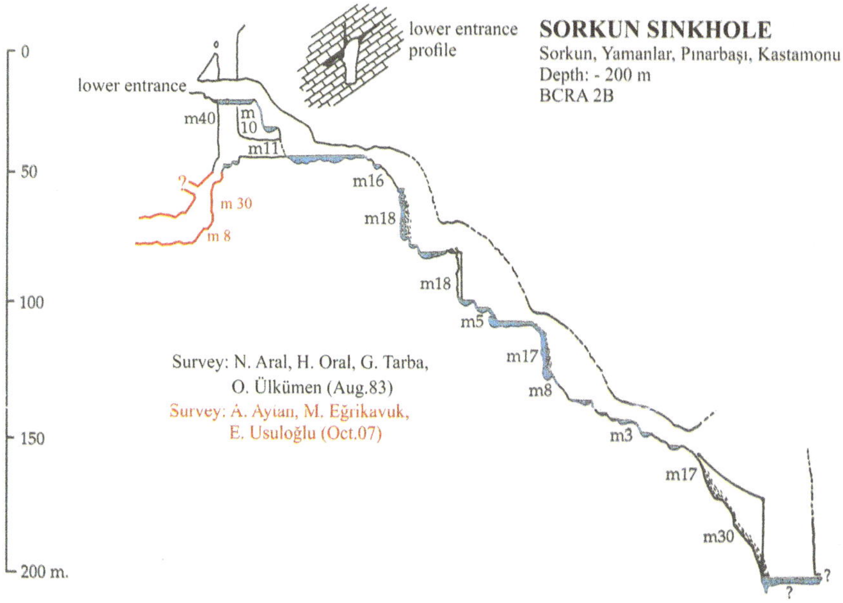

Fig. 3.88 Sorkun Sinkhole profile (drawn O. Ülkümen, E. Usuloğlu)

Fig. 3.89 Sorkun Sinkhole (photo E. Usuloğlu)

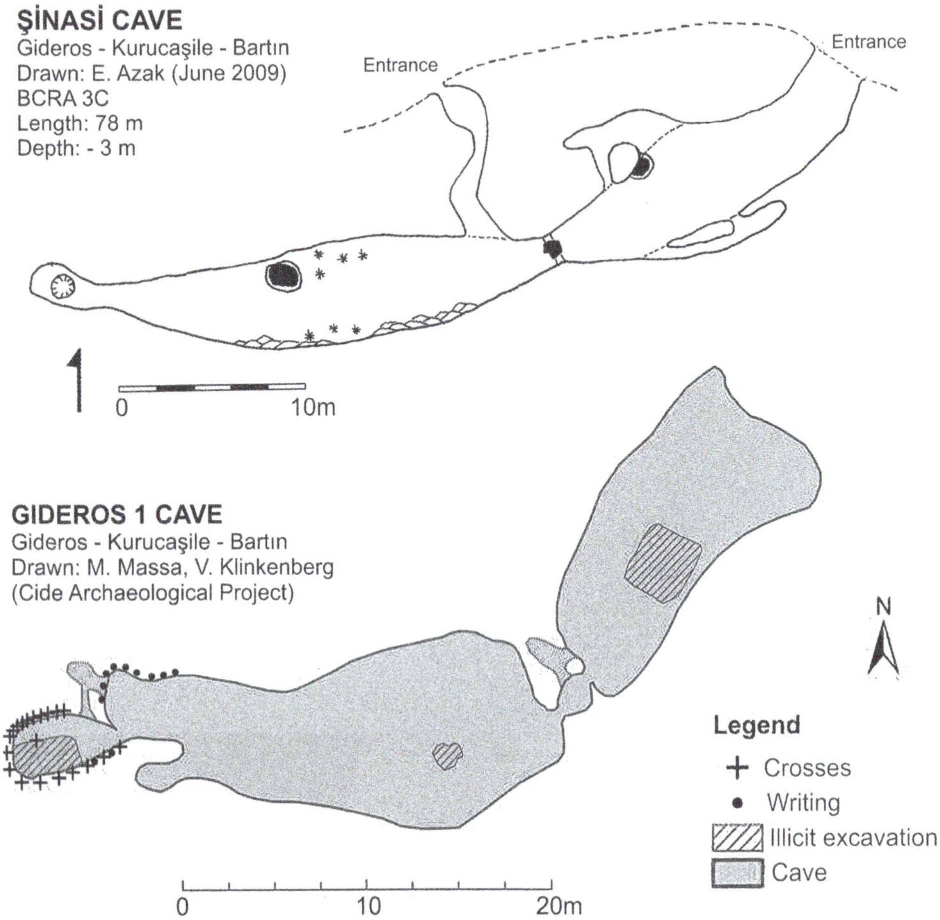

Fig. 3.90 Şinasi Cave (Gideros 1) double maps [drawn E. Azak (above) M. Massa, V. Klinkenberg (below)]

Fig. 3.91 Şinasi Cave (Gideros 1) (photo A. Yamaç)

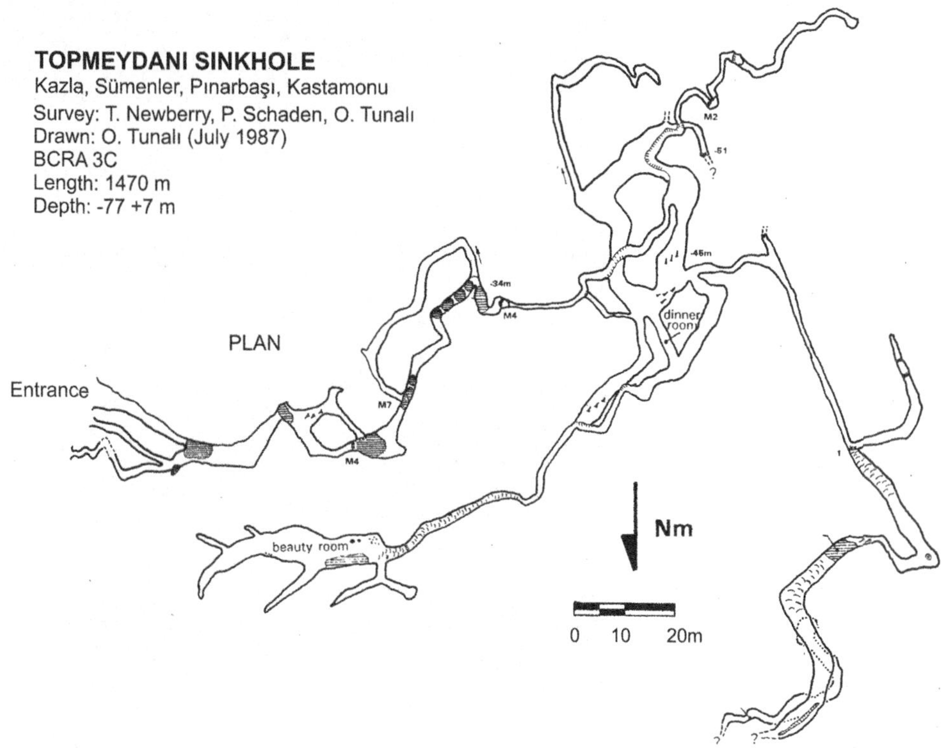

Fig. 3.92 Topmeydanı Sinkhole plan (drawn O. Tunalı)

Fig. 3.93 Topmeydanı Sinkhole (photo M. Albukrek)

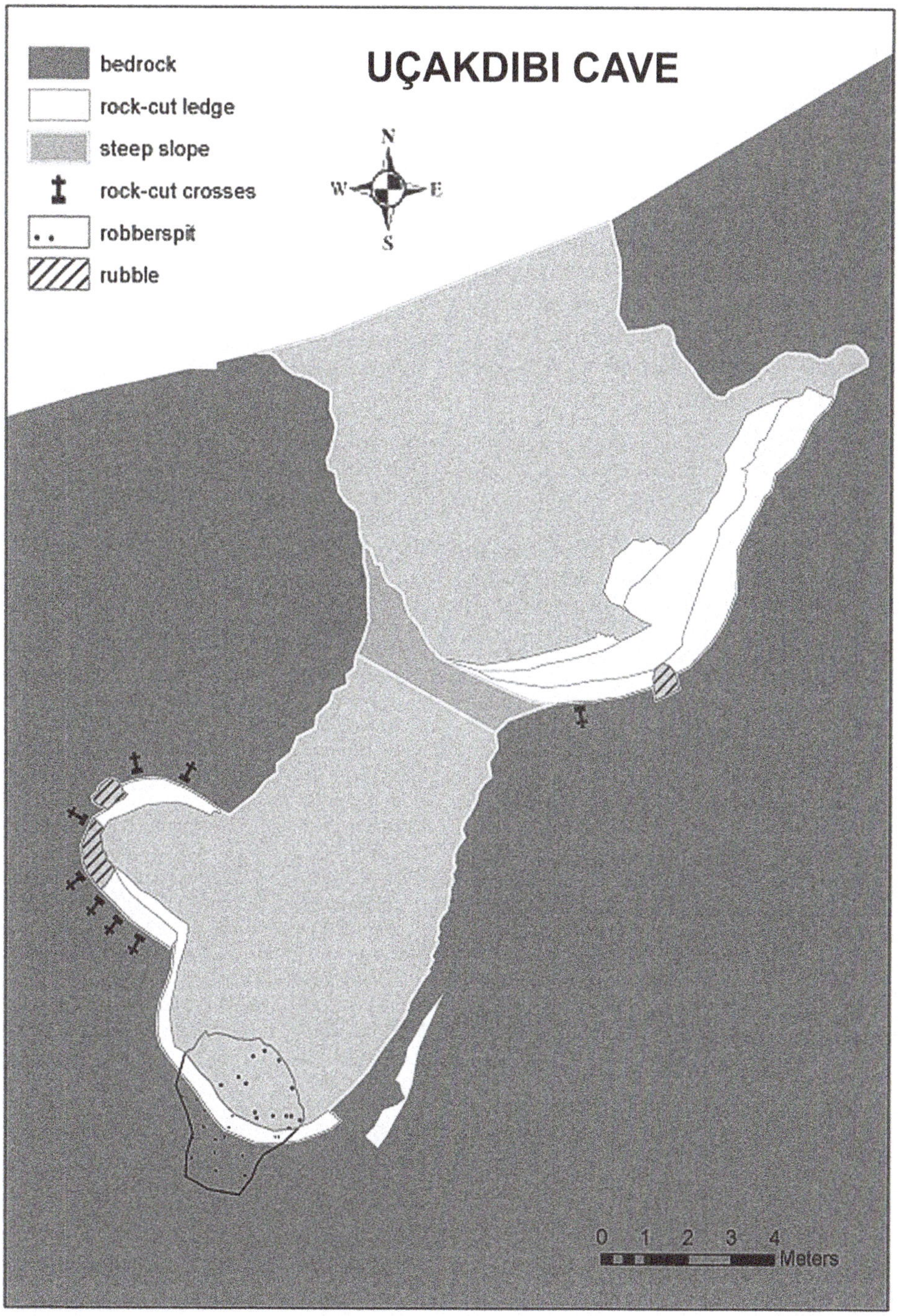

Fig. 3.94 Uçakdibi Cave plan (drawn Cide Archaeological Project)

Project conducted a survey of the area, and the report only briefly mentions the findings discovered in the cave.

Both Uçakdibi Cave and Gideros Cave 1 exhibit apse-like spaces. The apse at Uçakdibi is artificial, while that at Gideros is natural. In both caves, there are crosses incised or hewn into the rock. The cave at Uçakdibi is considerably more elaborate, comprising a large complex with artificially created ledges. The cave, which is a double cave, contains several drawn and carved crosses. Unfortunately, illegal excavation has disturbed the deposits, and the photos do not provide enough detail to date the cave or identify its occupant. The only Byzantine ceramics found are kitchen wares (Düring & Glatz, 2015) (Fig. 3.94).

References

Akkemik, Ü., Aytuğ, B., & Güzel, S. (2004). Archaeobotanical and Dendroarchaeological studies in Ilgarini Cave (Pınarbaşı, Kastamonu, Turkey). *Turkish Journal of Agriculture and Forestry, 28*(1), 9–17.

Akkemik, Ü., Aytuğ, B., & Güzel, S. (2013). Ilgarini Mağarası'nda Dendroarkeolojik Çalışmalar. *OBRUK, 6*.

Alagöz, C. (1944). *Türkiye Karst Olayları Hakkında Bir Araştırma (Une étude sur les phénomènes karstiques en Turquie)*. Ankara.

Albukrek, M. (1986). Kastamonu Bölgesi Araştırmaları II. *DELTA, 2*, 5–15.

Albukrek, M. (1993a). Kastamonu Bölgesi Araştırmaları III. *DELTA, 3*, 29–43.

Albukrek, M. (1993b). Kastamonu Bölgesi Araştırmaları IV. *DELTA, 4*, 15–25.

ASPEG. (2009). Küre Dağları Milli Parkı Mağara Biyoçeşitlilik Araştırma Projesi İkinci Ön Rapor (Unpublished Private Report).

Çelikoğlu, Ş., & Atış , E. (2015). Gürcüoluk Cave and it's Tourism Potential. *The Journal of International Social Research, 8*(41).

Dokay. (2018). *Bartın Mağaraları Hakkında Araştırma Raporu*. BAKKA Batı Karadeniz Kalkınma Ajansı.

Düring, B., & Glatz, C. (2015). *Kinetic landscapes the cide archaeological project: Surveying the Turkish Western Black Sea Region*. De Gruyter Open Ltd.

Düring, B., Glatz, C., & Şerifoğlu, T. (2012). The cide archaeological project 2011: Third preliminary report. *Anatolia Antiqua, 20*, 167–175.

Erdem, B. (1993). Karakaçak Mağarası. *DELTA, 6*, 55–57.

Glatz, C., Düring, B., Şerifoğlu, S., & Emre, T. (2011). The cide archaeological project 2010: Second preliminary report. *Anatolia Antiqua, 19*, 279–288.

Harris, S. (1992). FACC in Turkish delight. *Caves and Caving, 62*.

Holland, M. (1991). LUSS Turkey 90 expedition report. *Caves and Caving, 52*, 24–27.

Kuruçayırlı, E. (2004). Batı Karadeniz Bölgesi. *DELTA, 7*, 36.

Kuruçayırlı, E., & Osmanoğlu, B. (2004). Dağlı Kuylucu. *DELTA, 7*, 25–28.

LUSS. (1990). *Turkey '90*. Unpublished Report.

Nazik, L., Derici, Ş., Kutlay, L., Törk, K., Özel, E., Mengi, H., Aksoy, B., Acar, C., Güner, İ.N. (1999). *Gürcüoluk Mağarası (Amasra-Bartın) Mimari ve Elektrifikasyon Projesi*. Ankara: MTA Genel Müdürlüğü Jeoloji Etütleri Dairesi Başkanlığı.

Nazik, L., & Törk, K. (2003a). *Balıönü Mağarası (Cide - Kastamonu)*. MTA.

Nazik, L., & Törk, K. (2003b). *Ulus İlçesi Yakın Çevresi Doğal Mağaraları*. MTA.

Özbek, M. (2012). A new freshwater amphipod species, Gammarus obruki sp. nov., from Turkey (Amphipoda: Gammaridae). *Turkish Journal of Zoology TÜBİTAK, 36*(1), 567.

Törk, K., Özel, E., Acar, C., & Akgöz, M. (2008). *Kılıçlı Mağarası Araştırma Raporu*. Ankara MTA Genel Müdürlüğü.

Törk, K., Özel, E., & Akgöz, M. (2010). Kılıçlı Mağarası. *Mavi Gezegen*, 15.

Ülkümen, O. (1983). Kastamonu Bölgesi Araştırmaları I. *DELTA, 1*, 1–9.

Ünal, A. (1999). A hittite mycenaean type B sword from the vicinity of Kastamonu, Northwest Turkey. *Bulletin of the Middle Eastern Culture Center in Japan, XI*, 207–226.

Wolozan, D. (1992). *Recherches Spéléologiques en Turquie (4ème et 5ème campagnes de l'ADEKS)* (Vol. 3).

Yamaç, A. (2008). Bir Milli Parkta Kaybolmak. *OBRUK, 1*.

Yıldırım, H. (2008). Valla Kanyonu 23 Nisan Etkinliği. *MAD Bulletin, 14*, 119–123.

4 Biospeleology of Küre Mountains National Park

4.1 General Biospeleology

Biospeleology, also known as "Cave Biology," is a scientific branch that studies subterranean animals, their habitats, and their mutual relationships. The name comes from three words of Greek origin: *bios*, meaning life; *speleos*, meaning cavity; and *logos,* meaning science. This synthesis of sciences brings together two fundamental scientific disciplines: biology and speleology. The word biospeleology was first used by Armand Viré in his publication "*La biospéléologie*" in 1904 (Fig. 4.1). In it, he described the science which researches the living creatures inhabiting the underground. Also, he provided descriptions of many species and several new genera of cave inhabitants: Caecosphaeroma *burgundum*, Caecosphaeroma *faucheri*, Caecosphaeroma *virei*, Stenasellus *virea*, Niphargus *virei*, etc. Shortly after, biospeleology was formally established as a science following the publication of "Essai sur les problèmes biospéologiques," an article written by Emile Racovitza, a leading authority on the subject in Romania. Armand Viré and Emile Racovitza were accepted as the founders of biospeleology (Racovitza, 1907; Viré, 1904).

Caves are inhabited by a broad diversity of organisms—from microorganisms to large mammals—that are partly or entirely adapted to the demanding conditions of life underground, such as a lack of light, scarcity of food, and very high air humidity. Subterranean fauna refers to animal species that are adapted to live in an underground environment. To date, more than 7500 subterranean taxa (species and subspecies) have been described in the world. Beetles, crustaceans, molluscs, pseudoscorpions, spiders, and centipedes dominate the cave habitats of numerous species. These groups account for nearly 90% of all known cave taxa.

Cave dwellers are categorized into three groups as troglobites, troglophiles, and trogloxenes, according to the classification made by Racovitza in 1907. His classification is neither systematic nor phylogenetic; it is based solely on ecological considerations. Three main categories of cave dwellers are explained below, based on the classification of the dependence of cave organisms, i.e., cave troglofauna, on the cave environment during their life cycle.

Trogloxenes: These species occur only sporadically in an underground habitat and are unable to establish a subterranean population. These creatures, also called cave guests, are animal species that periodically live in underground habitats such as caves or at the entrance but cannot live exclusively in such habitats. They may require caves for a portion of their life cycle but must return to the surface for at least some part of their lives. For example, many creatures, such as bats, bears, rats, raccoons, birds, and some opiliones, can be found inside the caves. They spend certain parts of their lives, such as hibernation or nighttime, inside caves and depend on the outside for nourishment. Several extinct trogloxenes are known, such as the cave bear (Ursus spelaeus), cave lion, cave leopard, and cave hyena.

Troglophiles: These species live mainly above ground but also in subterranean habitats. Troglophiles can live part or all of their lives in caves but can also complete a life cycle in appropriate environments on the surface. They can be born in caves, grow, feed, reproduce, and die inside. These creatures can find and live in a cave similar to their outside environment. Examples are some species of salamanders, cave crickets, millipedes, pseudoscorpions, spiders, and some insect species.

Troglobites: These species are strictly bound to subterranean habitats. Troglobites are obligatory *cavernicoles* specialized for cave life. Some can leave caves for short periods and may complete parts of their life cycles above ground but cannot live their entire lives outside of a cave environment. Troglobites typically have evolutionary adaptations to cave life. Examples of such adaptations include slow metabolism, reduced energy consumption, better food usage efficiency, decrease or loss of eyesight, and depigmentation. Conversely, as opposed to lost or reduced functions, many species have evolved elongated antennae and locomotory appendages to move around better and respond to environmental stimuli.

A. Yamaç, E. Tok, *Caves and Karst of Küre Mountains National Park, Turkey*, Cave and Karst Systems of the World,
https://doi.org/10.1007/978-3-031-80142-6_4

Fig. 4.1 Armand Viré (Wikipedia)

These structures are also full of chemical, tactile, and humidity receptors. At the same time, these creatures have developed resistance to cold and lack nutrition for a long time, and some species can control their reproduction. For example, a species we call a ghostfish can keep its eggs in its mouth until conditions are suitable after spawning.

Troglobites commonly do not survive well outside caves and, therefore, cannot travel between separate cave systems. As a result, many troglobiotic species are endemic to a single cave or system of caves. Another feature of troglobites is that they are usually species-specific to their environment. They most likely entered and evolved in the narrow galleries of caves that began forming millions of years ago. The most up-to-date and probable theory on this subject is that the presence of these creatures under dense vegetation in sunless and humid places caused them to adapt to such environments, eventually choosing caves as their habitat and continuing to evolve there. Examples of troglobites include chemotrophic bacteria, some species of flatworms, velvet worms, pseudoscorpions, millipedes, centipedes, cavefish, and cave salamanders. The most famous member of this last group is *Proteus anguinus*.

In order to understand the ecosystem of the caves, it is helpful to know a little about the ecological regions of the cave. Light loses its effect from the entrance, with only reflected light remaining in the twilight zone. The region that follows is known as the transition zone. In this section, for the first time, there is complete darkness. The cave habitat is currently very reliant on the outside world, and is affected by variations in the weather, such as seasonal shifts. However, as we continue to the cave's depths, the zone where the light cannot penetrate begins. There is absolute darkness, and the temperature and humidity are nearly constant. These environmental constraints also affect the food pyramid in the caves, which is of vital importance for the ecological balance. Nutrients can enter the cave in a number of ways. Organic material such as humus, algae, etc., can be brought in by an incoming stream, if there is one, and larger sources of nutrition such as plants, animal remains, and even animals themselves can be brought in with seasonal floods. In addition, some material, such as pollen, leaves, and insects, can be carried in by the wind. The prey or the excrement of animals, such as bears, dormouse, and bats, can also be considered among the food sources observed widely within cave ecosystems.

Although largely isolated from the surface, caves should not be thought of as excluded ecosystems. To take the simplest example, if a cave is destroyed and the bats inside are removed from the area, they are most probably unfit to survive, as they are endemic to their habitats based on a fragile and delicate balance due to its extreme environmental conditions. This not only prevents the seeds of the plants from dispersing but also causes an increase in insects, some of which are agricultural pests, and this directly affects us by increasing the use of pesticides (Culver & Pipan, 2009; Romero, 2009). Besides, since hypogean habitats are associated not only with the troglofauna of terrestrial subterranean environments (caves and underground spaces above the water table), but also with the stygofauna of subterranean waters, the protection of life in caves includes not leaving them under dams, not blowing them up, and not using them as garbage dumps or natural sewers.

4.2 Biospeleology in Turkey

The first biospeleological research article in Turkey was written in 1899 by Ignacio Bolivar (1850–1944) (Fig. 4.2). Bolivar asked his friend M. Martinez Escalera, who went on a trip to Anatolia to study colepteres, to collect samples of some species for him during this trip. These insects, collected from the Kahramanmaraş and İskenderun caves, are the first cave creatures sampled in Anatolia. Ignacio Bolivar published his studies on the samples collected by his friend Escalera in the journal of the Belgian Entomology Society (Bolivar, 1899).

After this date, in the early 1930s, we see the research of G. Müller on Carabiden and P. de Beauchamp on Turbellaria.

Fig. 4.2 Ignacio Bolivar 1935 (Wikipedia)

Fig. 4.3 Curt Kosswig (Demirsoy, 2017)

From the second half of the 1930s onwards, the work of Curt Kosswig became a milestone not only in Anatolian biospeleology but also in the zoology of Turkey (Fig. 4.3). Kosswig, whose main field of study was the "Evolutionary biology of cave creatures" and who published many articles on this subject, guided many other scientists as well as his own work.

For example, Verhoeff, Strouhal, and Vandel studied woodlouse (Isopoda; Oniscidea) and millipede (Diplopoda); Jeannel studied beetles (Insecta; Coleoptera); and Beier studied pseudoscorpion (Arachnida; Pseudoscorpionida) (Jeannel, 1934a, 1934b, 1947; Verhoeff, 1936, 1940).

Mesoiulus kosswigi, Kosswigia insularis, Neobisium kosswigi, and *Trichonethes kosswigi* are just a few of the species collected by Kosswig from Anatolian caves and dedicated to his name by all these researchers (Demirsoy, 2017).

Another important name of biospeleology studies in Turkey is Knut Lindberg. Lindberg started his research in the caves around Zonguldak in 1952 and published many articles after these studies. In almost the same years, Coiffait was studying the rove beetle (Staphylinidae) in Southern Anatolia (Lindberg, 1952a, 1952b).

Since the second half of the 1960s, for the next 25 years, most of the biospeleological research in Anatolia has been carried out by Italian scientists, mostly from the University of Rome. Following the 1966, biospeleological field study in Anatolia by Valerio Sbordoni from the University of Rome, a total of 109 field studies were carried out with the participation of many famous zoologists from different disciplines, such as Augusto Vigna Taglianti, Paolo Marcello Brignoli, Achille Casale, and Marzio Zapparoli, and 197 scientific papers were published as a result of the evaluation of these studies (Brignoli, 1968, 1971, 1972; Casale, 1977, 1979; Casale & Giachino, 1985).

Diplocephalus turcicus, Dina vignai, Proasellus lykaonicus, Parhadzia sbordonii, and *Harpolithobius vignatagliantii* are some of the dozens of species that were sampled by Italian researchers from caves and groundwater in different regions of Anatolia and are new to the scientific world.

Beginning in the early 2000s, research on the invertebrates of Anatolian caves has been increasingly contributed by Turkish speleologists through several publications (Balık et al., 2002; Topçu et al., 2008; Yamaç et al., 2005) as well as academic theses (Paksuz, 2004; Selvi, 1999). In Turkey, a total of 203 cave-dwelling invertebrate species have been reported. Twenty-nine of the species are from the phylum Mollusca, 5 species are from the subclassis Oligochaeta, 1 species is from the subclassis Hirudinea, 82 species are from the classis Arachnida, 1 genus and 19 species are from the classis Diplopoda, 1 genus and 4 species are from the classis Chilopoda, 42 species are from the classis Insecta, and 21 species are from the subphylum Crustaceae. In these, 104 species are Anatolian endemics (Kunt et al., 2010).

Although not as long established as the study of cave-dwelling invertebrates, the study of Anatolian bat fauna and bat species databases began in the 1950s. The studies of pioneering researchers such as Şadoğlu, Kosswig, Eisentraut, and Strinati are noteworthy. For example, Şadoğlu's (1953) article on *R. aegyptiacus* is the first scientific publication on bats in Turkey (Şadoğlu, 1953). In the 1960s, species identification studies, which developed with the contributions of researchers such as Osborn, Spitzenberger, and Çağlar, brought along numerous scientific articles.

After 1990, although many different studies have been conducted on the subject (Furman & Özgül, 2004; Karataş, 2019), Petr Benda's bat species identification studies in Turkey and other countries of the Middle East are of particular importance. His database, "Bats (Mammalia: Chiroptera) of the Eastern Mediterranean," written in 16 different sections and covering all the countries from Albania to Libya, is more like an encyclopedia than a chain of articles. Part 1 of this work is the entire database of bats of Turkey (Benda & Horacek, 1998).

The number of recognized mammal species in the world is around 6,500, and after Rodentia, the most populous species is bats (Chiroptera). Bats are one of the most diverse and geographically dispersed orders of living mammals, with about 1,400 species on every continent. As far as has been discovered so far, 39 (one frugivorous and 38 insectivorous) bat species are distributed in Turkey.

4.3 Biospeleology of Küre Mountains National Park

The Küre Mountains National Park is a unique forest reserve area composed of a variety of habitats that support a diverse fauna. While the vegetation cover of the park consists of forest species dominated by pure or hybrid coniferous or deciduous trees, and parts occupied by herbaceous and bushy plants, this combination of different types of vegetation within the area results in a broad array of ecosystems in the Küre Mountains National Park.

The main objective of the project "Biospeleology of the Küre Mountains National Park" was to obtain a comprehensive understanding of the biological features of the subterranean part of this diverse ecosystem by investigating a number of caves that are related not only in terms of their geological and geographical features but also in terms of the climatic conditions to which they are exposed, distributed over a large area within the boundaries of the Küre Mountains National Park.

During this project, 200 biological samples were collected from 25 different caves in the Küre Mountains National Park between 2008 and 2010. The samples were analyzed by biologists, and 39 different taxa were identified at the genus or species level. In addition, nearly 85 taxa were systematically identified at the class and order level. Although some of the specimens collected had characteristics similar to those of the other caves found nearby, some other cave creatures were endemic to their environment. Preliminary evaluations of these examples are listed below.

A team of six carried out the biospeleological sampling and research in these 25 caves. We are indebted to Petar Beron, Milena Nikolova, Oana Chachula, Mehmet Sait Taylan, Ümit Kebapçı, and Özgür Can Sönmez for their help and efforts.

During the studies, standard biospeleological methods were used. Hydrobiological sampling was done with an Epstein plankton net and a hand net with a 100 μm diameter eye. The terrestrial fauna was sampled with exhausters, traps, or direct quantitative searching. The cave fauna in 25 caves was examined. Each cave had one sampling point at the entrance, one in the middle, and one at the end (or the end of the negotiable parts). For the deep sinkholes, 0 m, −30 m, −70 m, and −100 m (and −250 m for Sorkun) were sampled. An observer spent 5 min collecting on and adjacent to the transect across the cave at each site. The cave floor, walls, and roof were scanned, stones turned, guano searched, and penetrating tree roots examined. All the sampled invertebrates were preserved in 4% formaldehyde and 75% ethanol in vials. The organisms were separated in the laboratory and sent to taxonomic specialists for further determination. The vertebrates were enlisted only.

A considerable biodiversity of cave fauna was observed in the investigated section of the Küre Mountains. According to the primary assessment, 39 invertebrate taxa were found, and 14 may safely be presumed to be new species. The established biodiversity is present in Sect. 4.4.

Regarding the three groups of cave species, samples taken from different caves are mainly distributed as such:

Trogloxenes: Phalangiidae; Ephemeroptera—*Haproleptoides* sp.; *Ecdyonurus* sp.; Plecoptera, Culicidae; Muscidae; Lymnephilidae; Simuliidae *Crocidura leucodon*

Troglophiles: *Oxychilus* sp.; *Pisidium sp;* Astacidae; Collembola; Lymnobiidae; Pseudoscorpiones; *Triturus vittatus; Rana* cf. *temporaria;* Chiroptera

Troglobites: Carabidae; Niphargidae; Pseudoscorpiones; Diplopoda

According to this primary evaluation, the groups of Trogloxenes and Troglophiles were relatively equal. It can be expected that the number of species in the group of real troglobites will increase after the final determination of the samples. Similarly, the level of endemism of the cave fauna in the region will be finalized after new samples are taken.

The numbers of spiders, rosary beetles, and other insect species, including beetles, snails, and centipedes, were exceptionally high in the caves in the region. We encountered amphipods in almost all of the semi-active caves that contained water. While this number was high in horizontal caves, it was generally lower in vertical caves. Although there were large numbers of spiders at the entrances of the verticals, other species were in decline or absent.

According to the studies carried out so far, no significant bat population has been found in almost any cave except for a few bats (*Myotis myotis*) in horizontal caves such as Mantar, Atak, Medil 1, and Ilgarini. Experts stated that this situation is expected given the region's high altitude, which varies between 700 m and1000 m. On the other hand, mammals such as *Felis silvestris, Myoxus glis,* and *Glis glis* were also detected in the region, with a few skeletons found in Atak and İntürbesi caves, along with the remains of the food they consumed during their hibernation.

Küre Mountains National Park has a very high potential in terms of biodiversity in its caves, and the list of the creatures we documented in these caves can be seen on the following pages. Contrary to general belief, the high diversity of life at such a high altitude surprised us the most in this region.

The faunistic data identified by genus and species as a result of this project are given below in two separate sections as vertebrate and invertebrate fauna.

4.3.1 Vertebrate Fauna

From the samples collected within the scope of the Küre Mountains National Park Project, 4 taxa belonging to 2 different classes within the Vertebrata fauna were identified:

1. Mammalia: Within the scope of the project, the *Myotis myotis* taxon from the order Chiroptera, belonging to the class Mammalia, was identified.
2. Amphibia: As a result of the examination of the collected specimens, *Triturus* and *Salamandra salamandra* taxa, belonging to the Caudata order of the Amphibia class, and *Rana temproraria* species, belonging to the Anura order, were identified.

In addition to the Vertebrata fauna data given above, the skull and bone specimens of *Glis glis, Felis sylvestris,* and *Crocidura leucodon,* all belonging to the order Rodentia belonging to the class Mammalia, were identified after the investigations in the caves. These vertebrates, whose habitats are not caves, probably died by accidentally falling into the cave or the skulls or bones were carried into the cave.

4.3.2 Invertebrate Fauna

From the samples collected during the project, taxa belonging to the phyla Arthropoda, Mollusca, and Annelida were identified within the Invertebrata group. As a result of the research, two new species belonging to the orders Pseudoscorpionida and Amphipoda, two new species belonging to the classes Chilopoda and Diplopoda, and one new species belonging to the order Orthoptera—a total of five new species—were identified and brought to the stage of being presented as a publication by experts. If we group the taxa found during the project under the sub-heading of phyla:

1. Arthropoda: When the samples were examined, it was determined that they belonged to four different sub-branches of the Arthropoda branch. Four different taxa were found from the Chelicerata suborder. Three of them belong to the class Arachnida and are *Ixodes ricinus, Leiobunum,* and *Paranemastoma silli.* The other one belongs to the order Pseudoscorpionida and is thought to be a new species, and a publication on the taxon is being prepared.

   There are three taxa belonging to the Crustacea group, another suborder of the phylum Arthropoda. All three of them belong to the class Malacostraca and belong to the genera *Trichoniscus, Armadillidium,* and *Niphargus.* Of these taxa, the species belonging to the genus Niphargus is thought to be a new species, and experts are continuing their investigations to prepare a publication on this species.

   There were 12 taxa belonging to the class Insecta of the subfamily Mandibulata of Arthropoda. Two of these taxa belong to the genera *Lepismachilis* and *Machilus* of the order Thysanura, three taxa *Limonia nubelucossa, Culex pipiens,* and *Tarnaria* of the order Diptera, two taxa *Triphosa dubitata,* and *Scoleopteryx* of the order Lepidoptera. Two species of *Haproleptoides* and *Echyonurus,* belonging to the order Ephemeroptera; one species of *Carabus intricatus,* belonging to the order Coleoptera; and the last one of *Discoptila,* belonging to the order Orthoptera, as well as a new species belonging to the genus *Troglophilus* of the family Rhaphidophoridae, were discovered.

   As a result of their investigations, the biologists identified two new species belonging to the classes Chilopoda and Diplopoda of the suborder Myriapoda of the order Arthropoda. In addition, the genus *Lithobius* belonging to the class Chilopoda was identified.
2. Mollusca: From the samples collected from the caves examined within the scope of the Küre Mountains National Park project, two taxa were identified as belonging to the Mollusca. One of these taxa is *Oxychillus glaber* species belonging to the Gastropoda class, and the other is *Pisidium* genus belonging to Bivalvia class.
3. Annelida: Within the scope of the project, one taxon from Annelida was identified. This taxon is the genus *Hirudo* belonging to the class Clitellata.

4.3.3 Highlights of the Project

A total of six surveys were carried out in Atak Cave, and biological samples were collected in five of them. The different species found in these surveys were independently defined by biologists from Akdeniz and Burdur Universities. The unknown owner of the "nut shells," which are found in abundance throughout the cave, was one of our most pleasing discoveries. Dr. Ümit Kebapçı believes that the hibernating Dryomys nitedula or Glis glis is responsible for these mounds of nut shells, which are not found in the horizontal caves of the Topmeydan region, but have become a kind of fixture in the caves near Pınarbaşı.

Apart from a few small colonies of Myotis myotis in various locations, not many bats have been found despite the presence of considerable amounts of guano in some areas of the cave. On the other hand, the abundance of animal skeletons, especially near the entrances, is another dramatic illustration of the biodiversity of Atak Cave. A wildcat skull was easily identified in the first explorations. Interestingly, although specimens of at least five different spiders and centipedes were collected, not a single crustacean was found. Evidence of the presence of mammals was provided by a tick found on a wall about 60 meters from the cave entrance (Figs. 4.4, 4.5, and 4.6).

The dominant species in Buzluk and Dikilitaş Caves were several different families of Diptera. On the other hand, Döngelyanı and Ejder Caves had a rich diversity. The three species found in the samples from Ejder Cave, one of which is Pseudoscorpionida and the other two are Chilopoda and Diplopoda, may be new species.

Our research indicates that the Eşekçukuru Caves 1 and 2 have one of the highest biodiversity ratios in the region. Eşekçukuru Cave 2 has exceptionally intense biodiversity despite being an active sinkhole that absorbs a lot of water during peak season (Fig. 4.7). During the study, many specimens from both Chordata Vertebrata and Arthropoda Chelicerata were identified. Two of these specimens were published as new species (Andreev & Kenderov, 2012; Guéorguiev, 2012) (Figs. 4.8 and 4.9).

Ilgarini Cave is one of the most beautiful and important caves not only in Küre Mountains National Park but also in North West Anatolia. Although it is inside the National Park and not easily accessible, the increasing number of visitors every year causes ecological destruction in this cave. The biodiversity of the cave is diverse, and the interrelationships between these creatures are very complex. We are concerned that further increases in visitor numbers will disrupt this life cycle. Among 16 different specimens collected in this cave, Carabus *intricatus* and Limonia *nubeculosa* are remarkable (Figs. 4.10 and 4.11).

In the horizontal section just after the second descent of Inderesi Cave, several Gammarus specimens were collected,

Fig. 4.4 Atak Cave, Chilopoda Lithobiomorpha, *Lithobius sp.* (photo Ü. Kebapçı)

Fig. 4.5 Atak Cave, Dryomys (photo E. Soner)

Fig. 4.6 Atak Cave, Myotis myotis (photo A. Yamaç)

Fig. 4.7 Eşekçukuru Cave 2, *triturus sp.* (photo O. Chachula)

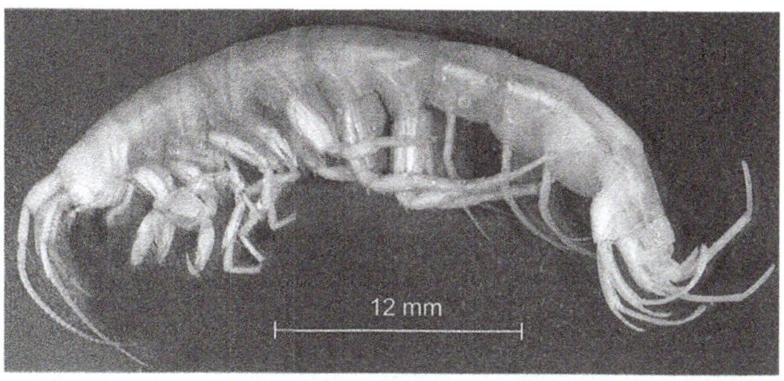

Fig. 4.8 Eşekçukuru Cave 2, Niphargus turcicus (photo S. Andreev)

Fig. 4.9 Eşekçukuru Cave 2, Beronaphaenops paphlagonicus (photo B.V. Gueorguiev)

and the samples were examined by Murat Özbek. As a result of the studies carried out in both genera, it was understood that it was a new species, and it was named after our caving group by Özbek: *Gammarus obruki* (Fig. 4.12) (Özbek, 2012).

Since Intürbesi Cave is a horizontal cave with low altitude and high humidity, it has a very diverse biodiversity spectrum. We can confidently state that it has a far richer biodiversity than many other caves in the area, despite the fact that we were only able to collect a small number of samples in just a brief period of time during our study. Two different examples—a Diptera and a Diplura—were particularly interesting (Figs. 4.13 and 4.14).

In Kapaklı Sinkhole, each sample was a different species. Among these samples, we found a *Niphargus*, which is possibly a new species. We encountered similar biodiversity in Kurtyolu 1 Sinkhole and Kuşkayası 1 Cave.

Despite its modest size, Mantar Cave was home to many different creatures. As a fossil cave, it was suitable for hibernation. Inside we found traces of spiders, beetles, and other insects; skeletons of bats; and traces of hibernating dryomys', among other species (Figs. 4.15 and 4.16).

A second unexpected cave ecosystem was discovered near the village of Makaracı; the Inderesi and Kocareis caves. In addition to a wide variety of snails and beetles, it was surprising to find two different species of amphipods coexisting in tiny pools above spider families.

In order to determine the ecological relationships between these organisms, various studies need to be designed by experts in the field.

A very rich life was discovered in Sorkun despite the fact that it is an active sinkhole and the highest flowing stream in the area is sinking into this cave. Although most of the species found in the cave belong to the subclassis Pterygota (Figs. 4.17 and 4.18), the only freshwater bivalvia encountered during the entire project was *Pisidium* sp., also found in Sorkun Sinkhole (Fig. 4.19).

Fig. 4.10 Ilgarini Cave, Carabus *intricatus* (photo O. Chachula)

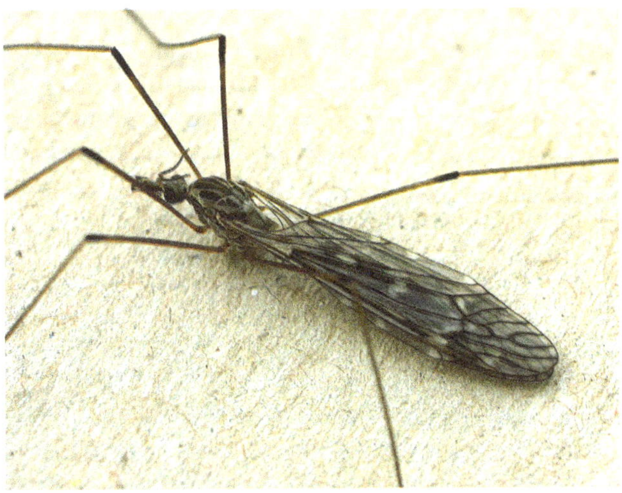

Fig. 4.11 Ilgarini Cave, Limonia *nubeculosa* (photo O. Chachula)

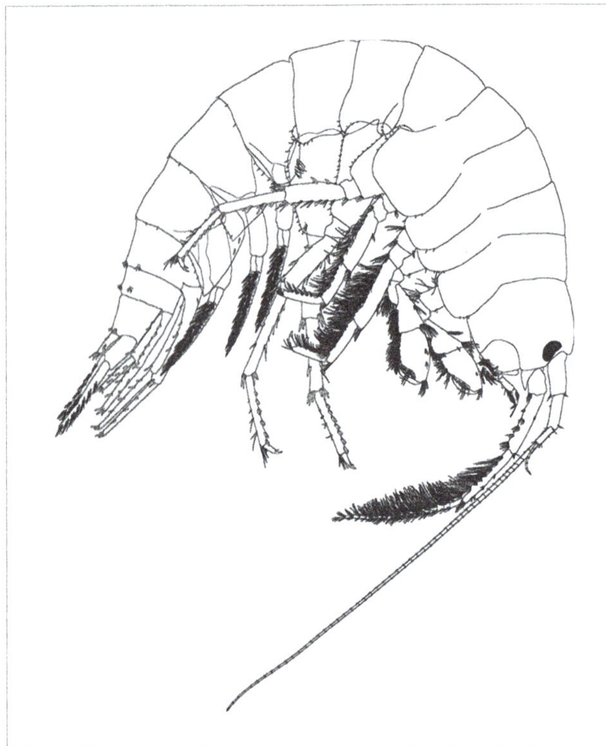

Fig. 4.12 Gammarus obruki sp. from Inderesi Cave (drawing M. Özbek)

Fig. 4.13 Intürbesi Cave, Diptera sciaridae (photo O. Chachula)

There were no significant finds in the Topmeydanı and Yukarı Döngelyanı sinkholes. Among the 21 samples taken from these two caves, the density was entirely Arthropoda, except for one or two samples. One of the few exceptions to this situation is an *Oxychilus* sp. discovered in both sinkholes.

In a short period of time, we have collected 200 different samples from 25 caves in the region. As a result of the analysis, 14 of these creatures were found to be new species that were previously unknown. We think that the number of

Fig. 4.14 Intürbesi Cave, Diplura (photo O. Chachula)

Fig. 4.16 Mantar Cave, diplopoda polydes (photo O. Chachula)

Fig. 4.15 Mantar Cave, zonitidae stereomicroscopical view (photo O. Chachula)

Fig. 4.17 Sorkun Sinkhole, Gastropoda—*Oxychilus* sp. (photo O. Chachula)

new species that will be discovered in unexplored caves and through more extensive studies is beyond any prediction.

We believe that all these caves with their vast biodiversity should be comprehensively protected, and their ecosystems must be studied before it is too late. Küre Mountains National Park is one of the 20 most important forest areas in the world, and its underground beauties are as magnificent and diverse as its aboveground beauties. We hope that future generations will also preserve and scrutinize this natural value with the same care.

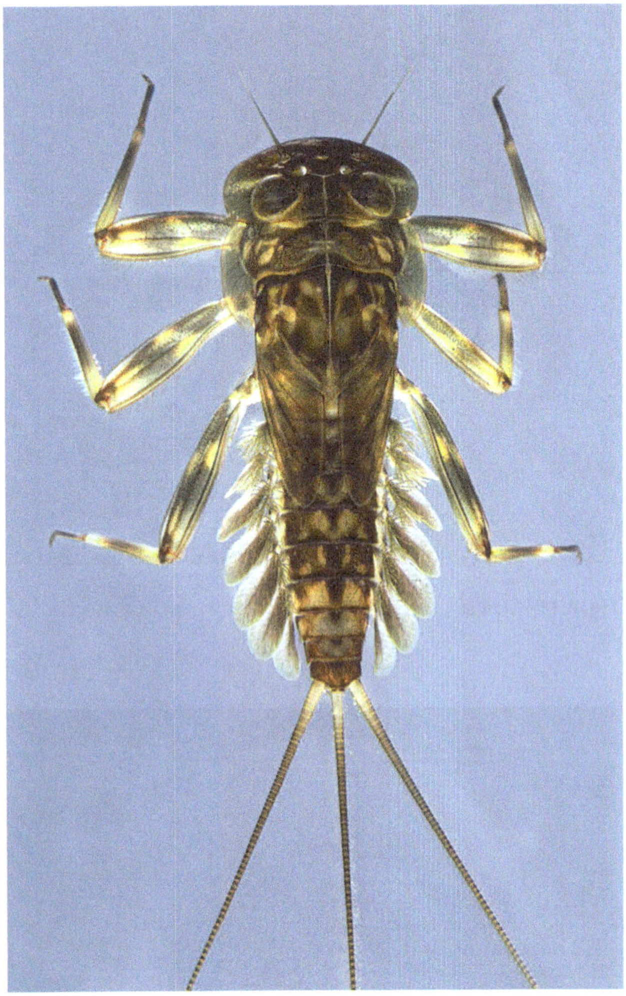

Fig. 4.18 Sorkun Sinkhole, Heptageniidae—*Echyonurus* sp. (photo O. Chachula)

Fig. 4.19 Sorkun Sinkhole, Bivalvia—*Pisidium* sp. (photo O. Chachula)

4.4 Taxonomic List of Küre Mountains National Park Cave Samples

Cave	Phylum	Subphylum	Classis	Subclassis	Ordo	Familia	Subfamilia	Genus	Species
Atak	Annelida		Clitellata	Hirudinea	Arhynchobdellida	Hirudinidae	Hirudinariinae	*Hirudo*	*Hirudo* sp.
Atak	Arthropoda	Myriapoda	Chilopoda	Anamorpha	Lithobiomorpha				
Atak	Arthropoda	Mandibulata	Insecta	Pterygota	Diptera				
Atak	Arthropoda	Chelicerata	Arachnida	Micrura	Araneae				
Atak	Arthropoda	Chelicerata	Arachnida	Acari	Ixodida	Ixodidae			
Atak	Arthropoda	Mandibulata	Insecta	Apterygota					
Atak	Arthropoda	Mandibulata	Insecta	Pterygota	Coleoptera				
Atak	Arthropoda	Chelicerata	Arachnida	Acari	Ixodida	Ixodidae	Ixodinae	*Ixodes*	*Ixodes ricinus* (male)
Atak	Arthropoda	Chelicerata	Arachnida	Dromopoda	Opiliones	Phalangiidae	Leiobuninae	*Leiobunum*	*Leiobunum sp*
Atak	Arthropoda	Chelicerata	Arachnida	Micrura	Araneae				
Atak	Arthropoda	Myriapoda	Chilopoda	Anamorpha	Lithobiomorpha	Lithobiidae	Lithobiinae	*Lithobius*	*Lithobius sp*
Atak	Arthropoda	Mandibulata	Insecta	Pterygota	Orthoptera				
Atak	Arthropoda	Mandibulata	Insecta	Pterygota	Lepidoptera				
Atak	Chordata	Vertebrata	Mammalia	Theria	Chiroptera	Vespertilionidae	Myotinae	*Myotis*	*Myotis* sp.
Atak	Chordata	Vertebrata	Mammalia	Theria	Rodentia	Gliridae	Glirinae	*Glis*	*Glis glis* (leg bone, nuts)
Atak	Chordata	Vertebrata	Mammalia	Theria	Carnivora	Felidae	Felinae	*Felis*	*Felis sylvestris* (skull)
Buzluk	Arthropoda	Mandibulata	Insecta	Pterygota	Diptera				
Buzluk	Arthropoda	Mandibulata	Insecta	Pterygota	Diptera				
Buzluk	Arthropoda	Mandibulata	Insecta	Pterygota	Diptera	Brachicera			
Buzluk	Arthropoda	Mandibulata	Insecta	Pterygota	Diptera	Limoniidae	Limoniinae	*Limonia*	*L. nubeculosa*
Buzluk	Chordata	Vertebrata	Mammalia	Theria	Chiroptera	Vespertilionidae	Myotinae	*Myotis*	*M. myotis*
Çöpbey	Arthropoda	Crustacea	Malacostraca	Eumalacostraca	Isopoda				
Dikilitaş	Arthropoda	Chelicerata	Arachnida	Micrura	Araneae				
Dikilitaş	Arthropoda	Chelicerata	Arachnida	Dromopoda	Opiliones				
Dikilitaş	Arthropoda	Crustacea	Malacostraca	Eumalacostraca	Isopoda				
Dikilitaş	Arthropoda	Mandibulata	Insecta	Apterygota	Collembola				
Dikilitaş	Arthropoda	Mandibulata	Insecta	Pterygota	Diptera	Limoniidae	Limoniinae	*Limonia*	*L. nubeculosa*
Dikilitaş	Arthropoda	Mandibulata	Insecta	Pterygota	Diptera	Limoniidae	Limoniinae	*Limonia*	*L. nubeculosa*
Dikilitaş	Arthropoda	Mandibulata	Insecta	Pterygota	Lepidoptera	Noctuidae		*Scoleopteryx*	*Scoleopteryx* sp.
Dikilitaş	Arthropoda	Mandibulata	Insecta	Pterygota	Trichoptera				
Dikilitaş	Mollusca	Conchifera	Gastropoda	Pulmonata	Stylommatophora	Zonitidae	Oxychilinae	*Oxychilus*	*Oxychilus* sp.
Dikilitaş	Arthropoda	Myriapoda	Diplopoda	Chilognatha	Luliformia				
Dikilitaş	Annelida		Clitellata	Oligochaeta					
Döngelyanı	Arthropoda	Chelicerata	Arachnida	Dromopoda	Opiliones				
Döngelyanı	Arthropoda	Chelicerata	Arachnida	Micrura	Araneae				
Döngelyanı	Arthropoda	Crustacea	Malacostraca	Eumalacostraca	Amphipoda				
Döngelyanı	Arthropoda	Crustacea	Malacostraca	Eumalacostraca	Isopoda				
Döngelyanı	Arthropoda	Mandibulata	Insecta	Pterygota	Coleoptera	Staphylinidae			
Döngelyanı	Arthropoda	Mandibulata	Insecta	Pterygota	Coleoptera				
Döngelyanı	Arthropoda	Mandibulata	Insecta	Pterygota	Coleoptera	Elateridae			
Döngelyanı	Arthropoda	Mandibulata	Insecta	Pterygota	Ephemeroptera				

Cave	Phylum	Subphylum	Classis	Subclassis	Ordo	Familia	Subfamilia	Genus	Species
Döngelyanı	Arthropoda	Mandibulata	Insecta	Pterygota	Trichoptera				
Döngelyanı	Arthropoda	Mandibulata	Insecta	Pterygota	Trichoptera				
Döngelyanı	Arthropoda	Mandibulata	Insecta	Pterygota	Diptera	Limoniidae	Limoniinae	*Limonia*	*L. nubeculosa*
Döngelyanı	Arthropoda	Mandibulata	Insecta	Pterygota	Diptera	Chironomidae			
Döngelyanı	Mollusca	Conchifera	Gastropoda	Pulmonata	Stylommatophora	Zonitidae	Oxychilinae	*Oxychilus*	*Oxychilus* sp.
Döngelyanı	Arthropoda	Myriapoda	Diplopoda	Chilognatha	Luliformia				
Döngelyanı	Annelida		Clitellata	Oligochaeta					
Ejder	Arthropoda	Chelicerata	Arachnida	Micrura	Araneae				
Ejder	Arthropoda	Chelicerata	Arachnida	Dromopoda	Pseudoscorpionida				*(possibly new* sp.*)*
Ejder	Arthropoda	Crustacea	Malacostraca	Eumalacostraca	Amphipoda				
Ejder	Arthropoda	Crustacea	Malacostraca	Eumalacostraca	Isopoda	Trichoniscidae	Trichoniscinae	*Trichoniscus*	*Trichoniscus* sp.
Ejder	Arthropoda	Mandibulata	Insecta	Pterygota	Coleoptera				
Ejder	Arthropoda	Mandibulata	Insecta	Pterygota	Diptera	Chironomidae			
Ejder	Arthropoda	Mandibulata	Insecta	Pterygota	Diptera	Culicidae	Culicinae	*Culex*	*C. pipiens*
Ejder	Arthropoda	Mandibulata	Insecta	Pterygota	Diptera	Limoniidae	Limoniinae	*Limonia*	*L. nubeculosa*
Ejder	Arthropoda	Mandibulata	Insecta	Apterygota	Thysanura	Machilidae	Machilinae	*Lepismachilis*	*Lepismachilis sp*
Ejder	Arthropoda	Mandibulata	Insecta	Pterygota	Trichoptera	Lymnephilidae			
Ejder	Arthropoda	Myriapoda	Chilopoda						*(possibly new* sp.*)*
Ejder	Arthropoda	Myriapoda	Diplopoda						*(possibly new* sp.*)*
Eşekçukuru 1	Chordata	Vertebrata	Amphibia	Lissamphibia	Caudata	Salamadridae	Pleurodelinae	*Triturus*	*Triturus* sp.
Eşekçukuru 1	Arthropoda	Chelicerata	Arachnida	Acari	Hydracarina				
Eşekçukuru 1	Arthropoda	Chelicerata	Arachnida	Micrura	Araneae				
Eşekçukuru 1	Arthropoda	Crustacea	Malacostraca	Eumalacostraca	Amphipoda	Niphargidae		*Niphargus*	*(possibly new* sp.*)*
Eşekçukuru 1	Arthropoda	Mandibulata	Insecta	Apterigota	Collembola				
Eşekçukuru 1	Arthropoda	Mandibulata	Insecta	Pterygota	Diptera	Limoniidae	Limoniinae	*Limonia*	*L. nubeculosa*
Eşekçukuru 1	Arthropoda	Mandibulata	Insecta	Pterygota	Diptera	Culicidae	Culicinae	*Culex*	*C. pipiens*
Eşekçukuru 1	Arthropoda	Mandibulata	Insecta	Pterygota	Trichoptera	Lymnephilidae			
Eşekçukuru 1	Arthropoda	Myriapoda	Chilopoda						*(possibly new* sp.*)*
Eşekçukuru 1	Arthropoda	Myriapoda	Diplopoda						*(possibly new* sp.*)*
Eşek Çukuru 2	Chordata	Vertebrata	Amphibia	Lissamphibia	Caudata	Salamandridae	Pleurodelinae	*Triturus*	*Triturus* sp.
Eşek Çukuru 2	Chordata	Vertebrata	Amphibia	Lissamphibia	Anura	Ranidae	Raninae	*Rana*	*R. temporaria*
Eşek Çukuru 2	Chordata	Vertebrata	Amphibia	Lissamphibia	Caudata	Salamandridae	Salamandrinae	*Salamandra*	*S. salamandra*
Eşek Çukuru 2	Arthropoda	Chelicerata	Arachnida	Micrura	Aranea				
Eşek Çukuru 2	Arthropoda	Chelicerata	Arachnida	Dromopoda	Opiliones				
Eşek Çukuru 2	Arthropoda	Chelicerata	Arachnida	Dromopoda	Opiliones				
Eşek Çukuru 2	Arthropoda	Crustacea	Malacostraca	Eumalacostraca	Decapoda				
Eşek Çukuru 2	Arthropoda	Crustacea	Malacostraca	Eumalacostraca	Isopoda				
Eşek Çukuru 2	Arthropoda	Crustacea	Malacostraca	Eumalacostraca	Isopoda	Trichoniscidae	Trichoniscinae	*Trichoniscus*	*Trichoniscus* sp.
Eşek Çukuru 2	Arthropoda	Crustacea	Malacostraca	Eumalacostraca	Amphipoda	Niphargidae			*(possibly new* sp.*)*
Eşek Çukuru 2	Arthropoda	Crustacea	Malacostraca	Eumalacostraca	Isopoda	Trichoniscidae			
Eşek Çukuru 2	Annelida		Clitellata	Hirudinea	Arhynchobdellida	Hirudinidae	Hirudinariinae	*Hirudo*	*Hirudo* sp.
Eşek Çukuru 2	Arthropoda	Mandibulata	Insecta	Apterigota	Thysanura	Machilidae	Machilinae	*Machilus*	*Machilus sp*
Eşek Çukuru 2	Arthropoda	Mandibulata	Insecta	Pterygota	Coleoptera				
Eşek Çukuru 2	Arthropoda	Mandibulata	Insecta	Apterigota	Collembola				

(continued)

(continued)

Cave	Phylum	Subphylum	Classis	Subclassis	Ordo	Familia	Subfamilia	Genus	Species
Eşek Çukuru 2	Arthropoda	Mandibulata	Insecta	Apterigota	Collembola				
Eşek Çukuru 2	Arthropoda	Mandibulata	Insecta	Pterygota	Diptera	Limoniidae	Limoniinae	*Limonia*	*L. nubeculosa*
Eşek Çukuru 2	Arthropoda	Mandibulata	Insecta	Pterygota	Diptera	Culicidae	Culicinae	*Culex*	*C. pipiens*
Eşek Çukuru 2	Arthropoda	Mandibulata	Insecta	Pterygota	Diptera	Mycetophylidae			
Eşek Çukuru 2	Arthropoda	Mandibulata	Insecta	Pterygota	Diptera	Chironomidae			
Eşek Çukuru 2	Arthropoda	Mandibulata	Insecta	Pterygota	Diptera	Culicidae	Culicinae	*Culex*	*C. pipiens*
Eşek Çukuru 2	Arthropoda	Mandibulata	Insecta	Pterygota	Diptera	Insecta			
Eşek Çukuru 2	Arthropoda	Mandibulata	Insecta	Pterygota	Lepidoptera	Geometridae	Larentiinae	*Triphosa*	*T. dubitata*
Eşek Çukuru 2	Arthropoda	Mandibulata	Insecta	Pterygota	Trichoptera	Lymnephilidae			
Ilgarini	Arthropoda	Mandibulata	Insecta	Pterygota	Diptera	Chironomidae			
Ilgarini	Arthropoda	Chelicerata	Arachnida	Dromopoda	Pseudoscorpionida				*(possibly new* sp.*)*
Ilgarini	Arthropoda	Crustacea	Malacostraca	Eumalacostraca	Amphipoda	Niphargidae		*Niphargus*	*(possibly new* sp.*)*
Ilgarini	Arthropoda	Crustacea	Malacostraca	Eumalacostraca	Isopoda				
Ilgarini	Arthropoda	Crustacea	Malacostraca	Eumalacostraca	Isopoda				
Ilgarini	Arthropoda	Mandibulata	Insecta	Pterygota	Coleoptera				
Ilgarini	Arthropoda	Mandibulata	Insecta	Pterygota	Coleoptera				
Ilgarini	Arthropoda	Mandibulata	Insecta	Pterygota	Coleoptera				
Ilgarini	Arthropoda	Mandibulata	Insecta	Pterygota	Coleoptera	Carabidae	Carabinae	*Carabus*	*C. intricatus*
Ilgarini	Arthropoda	Mandibulata	Insecta	Pterygota	Diptera	Culicidae	Culicinae	*Culex*	*C. pipiens*
Ilgarini	Arthropoda	Mandibulata	Insecta	Pterygota	Diptera	Limoniidae	Limoniinae	*Limonia*	*L. nubeculosa*
Ilgarini	Arthropoda	Mandibulata	Insecta	Pterygota	Diptera	Mycetophylidae			
Ilgarini	Arthropoda	Mandibulata	Insecta	Pterygota	Trichoptera	Lymnephilidae			
Ilgarini	Chordata	Vertebrata	Mammalia	Theria	Chiroptera	Vespertilionidae	Myotinae	*Myotis*	*Myotis* sp.
Ilgarini	Mollusca	Conchifera	Gastropoda	Pulmonata	Stylommatophora	Zonitidae	Oxychilinae	*Oxychilus*	*Oxychilus* sp.
Ilgarini	Arthropoda	Myriapoda	Chilopoda/Symphyla						
Ilgarini	Arthropoda	Myriapoda	Diplopoda						
İnderesi	Arthropoda	Crustacea	Malacostraca	Eumalacostraca	Amphipoda	Gammaridae	Gammarus		*G. obruki* sp. *(new* sp.*)*
İntürbesi	Arthropoda	Myriapoda	Diplopoda	Chilognatha	Julida				
İntürbesi	Arthropoda	Mandibulata	Insecta	Apterygota	Diplura				
İntürbesi	Arthropoda	Mandibulata	Insecta	Pterygota	Diptera	Sciaridae			
İntürbesi	Chordata	Vertebrata	Mammalia	Theria	Rodentia	Gliridae	Glirinae	*Glis*	*Glis glis* (nuts)
Kapaklı	Arthropoda	Crustacea	Malacostraca	Eumalacostraca	Amphipoda	Niphargidae		*Niphargus*	*(possibly new* sp.*)*
Kapaklı	Arthropoda	Crustacea	Malacostraca	Eumalacostraca	Isopoda	Onisicidea			
Kapaklı	Arthropoda	Crustacea	Malacostraca	Eumalacostraca	Isopoda	Trichoniscidae	Trichoniscinae	*Trichoniscus*	*Trichoniscus* sp.
Kapaklı	Arthropoda	Mandibulata	Insecta	Pterygota	Diptera	Culicidae	Culicinae	*Culex*	*C. pipiens*
Kapaklı	Arthropoda	Mandibulata	Insecta	Pterygota	Diptera	Limoniidae	Limoniinae	*Limonia*	*L. nubeculosa*
Kapaklı	Arthropoda	Mandibulata	Insecta	Pterygota	Trichoptera	Lymnephilidae			
Kapaklı	Chordata	Vertebrata	Mammalia	Theria	Soricomorpha	Soricidae	Crocidurinae	*Crocidura*	*C. leucodon*
Kaval Cıngırı	Arthropoda	Chelicerata	Arachnida	Micrura	Araneae				
Kocareis	Arthropoda	Crustacea	Malacostraca	Eumalacostraca	Isopoda				
Kocareis	Chordata	Vertebrata	Mammalia	Theria	Chiroptera	Vespertilionidae	Myotinae	*Myotis*	*Myotis* sp.
Kocareis	Arthropoda	Chelicerata	Arachnida	Dromopoda	Opiliones				
Kocareis	Mollusca	Conchifera	Gastropoda						
Kurtyolu 1	Arthropoda	Crustacea	Malacostraca	Eumalacostraca	Amphipoda				

(continued)

Cave	Phylum	Subphylum	Classis	Subclassis	Ordo	Familia	Subfamilia	Genus	Species
Kurtyolu 1	Mollusca	Conchifera	Gastropoda						
Kurtyolu 1	Arthropoda	Myriapoda	Chilopoda						
Kurtyolu 1	Arthropoda	Crustacea	Malacostraca	Eumalacostraca	Isopoda				
Kurtyolu 1	Annelida		Clitellata	Hirudinea	Arhynchobdellida	Hirudinidae	Hirudinariinae	*Hirudo*	*Hirudo* sp.
Kurtyolu 1	Arthropoda	Mandibulata	Insecta	Pterygota	Lepidoptera				
Kurtyolu 1	Arthropoda	Mandibulata	Insecta	Pterygota	Diptera				
Kurtyolu 1	Arthropoda	Mandibulata	Insecta	Pterygota	Coleoptera				
Kurtyolu 1	Arthropoda	Chelicerata	Arachnida	Micrura	Araneae				
Kurtyolu 1	Arthropoda	Mandibulata	Insecta	Pterygota	Orthoptera	Gryllidae	Gryllomoprhinae	*Discoptila*	*Discoptila* sp.
Kuşkayası 1	Arthropoda	Myriapoda	Chilopoda						
Kuşkayası 1	Arthropoda	Mandibulata	Insecta	Pterygota	Hymenoptera				
Kuşkayası 1	Arthropoda	Crustacea	Malacostraca	Eumalacostraca	Isopoda				
Kuşkayası 1	Arthropoda	Chelicerata	Arachnida	Dromopoda	Pseudoscorpionida				
Mantar	Arthropoda	Chelicerata	Arachnida	Dromopoda	Opiliones			*Paranemastoma*	*P. silii*
Mantar	Arthropoda	Crustacea	Malacostraca	Eumalacostraca	Isopoda	Trichoniscidae	Trichoniscinae	*Trichoniscus*	*Trichoniscus* sp.
Mantar	Arthropoda	Mandibulata	Insecta	Apterygota	Collembola				
Mantar	Arthropoda	Mandibulata	Insecta	Pterygota	Diptera	Chironomidae			
Mantar	Arthropoda	Mandibulata	Insecta	Pterygota	Diptera	Culicidae	Culicinae	*Culex*	*C. pipiens*
Mantar	Arthropoda	Mandibulata	Insecta	Pterygota	Diptera	Limoniidae	Limoniinae	*Limonia*	*L. nubeculosa*
Mantar	Arthropoda	Mandibulata	Insecta	Pterygota	Trichoptera	Lymnephilidae			
Mantar	Chordata	Vertebrata	Mammalia	Theria	Chiroptera	Vespertilionidae	Myotinae	*Myotis*	*M. myotis*
Mantar	Mollusca	Conchifera	Gastropoda	Pulmonata	Stylommatophora	Zonitidae	Oxychilinae	*Oxychilus*	*Oxychilus* sp.
Mantar	Mollusca	Conchifera	Gastropoda	Pulmonata	Stylommatophora	Zonitidae	Oxychilinae	*Oxichylus*	*Oxychilus* sp.
Mantar	Arthropoda	Myriapoda	Chilopoda						(*possibly new* sp.)
Mantar	Arthropoda	Myriapoda	Diplopoda						(*possibly new* sp.)
Medil 1	Arthropoda	Chelicerata	Arachnida	Micrura	Araneae				
Medil 1	Arthropoda	Crustacea	Malacostraca	Eumalacostraca		Armadillidiidae		*Armadillidium*	*Armadillidium* sp.
Medil 2	Annelida		Clitellata	Hirudinea	Arhynchobdellida	Hirudinidae	Hirudinariinae	*Hirudo*	*Hirudo* sp.
Sipahiler	Chordata	Vertebrata	Mammalia	Theria	Chiroptera				
Sipahiler	Arthropoda	Mandibulata	Insecta	Pterygota	Orthoptera	Rhaphidophoridae	Troglophlinae	*Troglophilus*	*Troglophilus sp (new sp)*
Sipahiler	Arthropoda	Mandibulata	Insecta	Pterygota	Diptera	Culicidae			
Sipahiler	Arthropoda	Mandibulata	Insecta	Pterygota	Diptera	Limoniidae			
Sipahiler	Arthropoda	Chelicerata	Arachnida	Micrura	Araneae				*sp1*
Sipahiler	Arthropoda	Chelicerata	Arachnida	Micrura	Araneae				*sp2*
Sipahiler	Arthropoda	Chelicerata	Arachnida	Micrura	Araneae				*sp3*
Sipahiler	Arthropoda	Crustacea	Malacostraca	Eumalacostraca	Isopoda				
Sipahiler	Arthropoda	Myriapoda	Diplopoda						
Sipahiler	Mollusca	Conchifera	Gastropoda						
Sorkun	Arthropoda	Chelicerata	Arachnida	Acari	Hydracarina				
Sorkun	Arthropoda	Crustacea	Malacostraca	Eumalacostraca	Amphipoda	Niphargidae		*Niphargus*	(*possibly new* sp.)
Sorkun	Arthropoda	Crustacea	Branchiopoda	Phyllopoda	Cladocera				
Sorkun	Arthropoda	Crustacea	Maxillopoda	Copepoda					
Sorkun	Arthropoda	Crustacea	Malacostraca	Eumalacostraca	Decapoda	Astacidae			
Sorkun	Arthropoda	Crustacea	Ostracoda						

(continued)

(continued)

Cave	Phylum	Subphylum	Classis	Subclassis	Ordo	Familia	Subfamilia	Genus	Species
Sorkun	Arthropoda	Mandibulata	Insecta	Pterygota	Diptera				
Sorkun	Arthropoda	Mandibulata	Insecta	Pterygota	Diptera	Simuliidae			
Sorkun	Arthropoda	Mandibulata	Insecta	Pterygota	Diptera	Culicidae	Culicinae	*Culex*	*C. pipiens*
Sorkun	Arthropoda	Mandibulata	Insecta	Pterygota	Diptera	Limoniidae	Limoniinae	*Limonia*	*L. nubeculosa*
Sorkun	Arthropoda	Mandibulata	Insecta	Pterygota	Ephemeroptera	Leptophlebiidae		*Haproleptoides*	*Haproleptoides* sp.
Sorkun	Arthropoda	Mandibulata	Insecta	Pterygota	Ephemeroptera	Heptageniidae		*Echyonurus*	*Echyonurus* sp.
Sorkun	Arthropoda	Mandibulata	Insecta	Pterygota	Plecoptera				
Sorkun	Arthropoda	Mandibulata	Insecta	Pterygota	Trichoptera	Lymnephilidae			
Sorkun	Mollusca		Bivalvia	Metabranchia	Veneroida	Sphaeriidae		*Pisidium*	*Pisidium* sp.
Sorkun	Mollusca	Conchifera	Gastropoda	Pulmonata	Stylommatophora	Zonitidae	Oxychilinae	*Oxychilus*	*Oxychilus* sp.
Sorkun	Arthropoda	Myriapoda	Diplopoda						
Şelale	Arthropoda	Chelicerata	Arachnida	Micrura	Araneae				
Şelale	Arthropoda	Chelicerata	Arachnida	Dromopoda	Opiliones				
Topmeydanı	Nematoda								
Topmeydanı	Arthropoda	Mandibulata	Insecta	Pterygota	Diptera	Culicidae	Culicinae	*Culex*	*C. pipiens*
Topmeydanı	Arthropoda	Mandibulata	Insecta	Pterygota	Diptera	Limoniidae	Limoniinae	*Limonia*	*L. nubeculosa*
Topmeydanı	Arthropoda	Mandibulata	Insecta	Pterygota	Trichoptera	Lymnephilidae			
Topmeydanı	Arthropoda	Chelicerata	Arachnida	Micrura	Araneae				
Topmeydanı	Arthropoda	Crustacea	Malacostraca	Eumalacostraca	Isopoda	Asellota			
Topmeydanı	Arthropoda	Crustacea	Ostracoda						
Topmeydanı	Arthropoda	Mandibulata	Insecta	Pterygota	Odonata				
Topmeydanı	Mollusca	Conchifera	Gastropoda	Pulmonata	Stylommatophora	Zonitidae	Oxychilinae	*Oxychilus*	*Oxychilus* sp.
Topmeydanı	Annelida		Clitellata	Oligochaeta					
Yukarı Döngelyanı	Arthropoda	Mandibulata	Insecta	Pterygota	Diptera	Mycetophylidae		*Tarnaria*	*Tarnaria* sp.
Yukarı Döngelyanı	Arthropoda	Mandibulata	Insecta	Pterygota	Diptera	Limoniidae	Limoniinae	*Limonia*	*L. Nubeculosa*
Yukarı Döngelyanı	Arthropoda	Mandibulata	Insecta	Pterygota	Trichoptera				
Yukarı Döngelyanı	Mollusca	Conchifera	Gastropoda	Pulmonata	Stylommatophora	Zonitidae	Oxychilinae	*Oxychilus*	*Oxychilus* sp.

References

Andreev, S., & Kenderov, L. (2012). Sur une nouvelle espèce du genre Niphargus de la Turquie—Niphargus turcicus n.sp. (Amphipoda, Niphargidae). *Historia Naturalis Bulgarica, 20*, 47–56.

Balık, S., Ustaoğlu, R., Özbek, M., Taşdemir, A., & Topkara, E. (2002). Yelköprü Mağarası (Dikili, İzmir) ve Yakın Çevresinin Sucul Faunası Hakkında Bir Ön Araştırma. *E.U. Journal of Fisheries & Aquatic Sciences, 19-1*(2), 221–225.

Benda, P., & Horacek, I. (1998). Bats of the Eastern Mediterranean. Part 1. Review of distribution and taxonomy of bats in Turkey. *Acta Societatis Zoologicae Bohemicae, 62*, 255–313.

Bolivar, I. (1899). Orthoptères du voyage de M. Martinez Escalera dans l'Asie Mineure. *Bulletin et Annales de la Societe Etomologique de Belgique*, 43.

Brignoli, P. (1968). Due nuove Paraleptoneta cavernicole dell'Asia Minore (Aranae, Leptonetidae). *Fragmenta Entomologica, 6*, 23–37.

Brignoli, P. (1971). Un nuoveo Troglohyphantes cavernicolo ed anoftalmo (ouanophthalmo) dell'Asia minore (Araneae, Linyphiidae). *Fragmenta Entomologica, 7*(2), 73–77.

Brignoli, P. (1972). Terzo contributo alla conoscenza dei Ragni cavernicoli di Turchia (Araneae). *Fragmenta Entomologica, 8*(3), 161–190.

Casale, A. (1977). Nuova Specie di Pterostichidae della Turchia (Coleoptera). *Fragmenta Entomologica, 13*, 119–124.

Casale, A. (1979). Un Nuovo Duvalius dell'Armenia Turca (Coleoptera, Carabidae, Trechinae). *Bollettino del Museo di Zoologia Dell'universita di Torino, 7*, 145–152.

Casale, A., & Giachino, P. (1985). Nuovi Bathyscinae (Coleoptera, Catopidae) di Grecia e di Turchia. *Bollettino Museo Regionale di Scienze Naturali, 3*(1), 221–231.

Culver, D., & Pipan, T. (2009). *The biology of caves and other subterranean habitats*. Oxford Oxford University Press.

Demirsoy, A. (2017). *Rüyalarını Türkçe Gören Bir Alman Bilim Adamı: Ord. Prof. Dr. Curt Kosswig*. Palme Yayıncılık.

Furman, A., & Özgül, A. (2004). The distribution of cave-dwelling bats and conservation status of underground habitats in Northwestern Turkey. *Biological Conservation, 120*(2), 243–248.

Guéorguiev, B. (2012). Beronaphaenops paphlagonicus, a new anophthalmous genus and species of Trechini (Coleoptera, Carabidae) from Turkey. *ZooKeys, 255*, 73–92.

Jeannel, R. (1934a). Coléoptères Cavernicoles de la Grotte de Fersine en Asie Mineure. *Annales Société Entomologique de France, 3*, 159–174, 345–346.

Jeannel, R. (1934b). Diagnoses préliminaires de deux coléoptères cavernicoles de l'Asie Mineure. *Bulletin Société Entomologique de France*, 20–21.

Jeannel, R. (1947). Coleopteres cavernicoles de l'Anatolie recueillis par M.C. Kosswig. *Notes Biospéologiques, 1*, 9–15.

Karataş, A. (2019). The bats (Mammalia: Chiroptera) of the Central and Eastern Mediterranean Region, with description of two new subspecies. *Acta Biologica Turcica, 32*(1), 42–52.

Kunt, K., Yağmur, E., Özkütük, S., Durmuş, H., & Anlaş, S. (2010). Checklist of the Cave Dwelling Invertebrates (Animalia) of Turkey. *Biological Diversity and Conservation, 3/2* 26–41.

Lindberg, K. (1952a). Notes sur Quelques Grottes de la Turquie avec Liste des Cyclopides (crustacés copépodes) recueillis dans ces grottes. *Annales de Spéléologie, VII*(1), 55–66.

Lindberg, K. (1952b). Un nouveau cyclopide (Crustace copépode) cavernicole de la Turquie d' Europe. *Bulletin de la Société Zoologique de France, LXXVII*, 437–440.

Özbek, M. (2012). A new freshwater amphipod species, Gammarus obruki sp. nov., from Turkey (Amphipoda: Gammaridae). *Turkish Journal of Zoology TÜBİTAK, 36*(1).

Paksuz, S. (2004). *Dupnisa mağara sistemi yarasaları (Mammalia; Chiroptera)*. PhD, Trakya Üniversitesi, Unpublished.

Racovitza, E. (1907). Essai sur les problèmes biospéologiques. *Biospéologica, 1* (Archives de zoologie expérimentale et générale), 371–488.

Romero, A. (2009). *Cave biology, life in darkness*. Cambridge University Press.

Şadoğlu, P. (1953). Meyva yiyen yarasalarin beslenmesi [Feeding of the fruit-eating bats]. *Biologi, 3*, 12–17.

Selvi, B. (1999). *Ballıca Mağarası (Tokat) Alg Florası*. PhD, Gaziosmanpaşa Üniversitesi, Unpublished.

Topçu, A., Türkeş, T., & Seyyar, O. (2008). A new spider of the genus Troglohyphantes (Araneae: Linyphiidae) from a Turkish cave. *Zoology in the Middle East, 45*, 91–95.

Verhoeff, K. (1936). Uber einige Myriapoden und einen Isopoden aus Mediterranen Höhlen. *Mitteilungen uber Höhlen und Karstforschung*, 155–162.

Verhoeff, K. (1940). Über Diplopoden aus der Türkei. *İstanbul Üniversitesi Fen Fakültesi Mecmuası, 5*(1–2), 1–49.

Viré, A. (1904). La biospéléologie. *Comptes rendus de l'Académie des sciences, CXXXIX*, 992–995.

Yamaç, M., Erginbaş, G., Şahin, N., Işık, K., Karaduman, A., Aslan, C., … Çakmak, G. (2005). *Some biological activities of Actinomycete isolates from Ayvacık Düdeni cave (Ödemiş-Izmir)*. Paper presented at the National Cave Days Symposium, Ankara.

5 Conclusion

Küre Mountains National Park represents one of the most significant natural assets of Turkey. In addition to its geological and geomorphological significance, the park boasts a rich biodiversity, including a diverse array of flora and fauna.

With 125 caves that have already been explored, this national park has the most caves of any national park in the world. As previously stated in Chap. 3, the number of caves in the region is likely to increase exponentially in the future.

On the other hand, scientific investigations of these caves have only recently commenced. The cave archaeology conducted in a limited area and the biospeleological research carried out by us for 3 years have been the subject of articles with findings of great interest. For instance, our research has led to the discovery and publication of 15 previously unknown species. It is currently impossible to predict the results of future scientific studies in these caves.

Küre Mountains National Park is home to 40 of the 132 mammal species known in Anatolia, as well as 129 bird species from 38 families. It serves as a sanctuary for numerous endangered species. While limited information is available on mammals and birds, knowledge regarding the area's reptiles and amphibians remains insufficient.

Additionally, the park is notable for its significant endemism of plants. Recent studies have identified 250 endemic plant taxa within the park's boundaries, which represents nearly 60% of all endemic plant taxa in Turkey (Tunçkol et al., 2020, 2022).

It may appear paradoxical and embarrassing to discuss the protection of a national park that has such unique and priceless components. In essence, the primary and overarching objective of a national park is to ensure its protection. In other words, it is imperative that all living creatures, plants, and natural structures within a national park are safeguarded from any external factors that might cause them harm and that they are permitted to continue their existence in the same manner.

It's not! Because, the laws and regulations can easily be bent by high-level government officials. While the rivers in the region are subject to hydroelectric power plant projects, another environmental issue is the construction of glass terraces and walking paths in the canyons, as well as the unplanned opening of caves to tourism. This phenomenon has become commonplace in Turkey, largely due to officials who lack vision and creativity and copy models seen elsewhere to generate simple advertising or tourist revenue.

Horma and Çatak canyons have been particularly affected by this phenomenon. These structures, which cannot be compensated for the damage they have caused to nature, are observed not only in Küre Mountains National Park but also in almost every part of Turkey.

In 1981, we spent 7 h searching for the Ilgarini Cave in the forest. However, today the same cave has a path, hundreds of visitors, and tables where they can rest in front of the cave. It is difficult to imagine the level of ecological damage that new practices in the future might cause. It is important to consider the impact of these arrangements on the region's biodiversity. Many endemic plants, some of which may yet be discovered, are disappearing due to the construction of these facilities. Such developments are incompatible with the protection of the Küre Mountains National Park, which should be completely preserved.

A number of scientists who conduct research in Küre Mountains National Park, specializing either in flora or fauna, have published articles in which they express concern about the impact of these constructions on endemic species in the region (Andreev & Kenderov, 2012; Guéorguiev, 2012; Tunçkol et al., 2020).

Another significant concern is the protection of several crucial caves situated within the buffer zone of Küre Mountains National Park, which lack the designation of national park protection status. A number of caves, including Atak, Buzluk, and Sipahiler, possess significant archaeological, biological, and structural value. However, these caves remain vulnerable to various forms of destruction.

On the other hand, we are aware that it is naive to hope for a protection project for the caves in the buffer zone when

A. Yamaç, E. Tok, *Caves and Karst of Küre Mountains National Park, Turkey*, Cave and Karst Systems of the World,
https://doi.org/10.1007/978-3-031-80142-6_5

even the caves within the borders of the national park are not adequately protected. Regrettably, in Turkey, nature is still regarded as a commercial commodity, and protection is only implemented when there is a commercial incentive. In the absence of such an incentive, we frequently observe the exploitation of caves for the purposes of mining or quarrying. Consequently, despite our initial optimism and aspirations expressed at the outset of this publication, we believe that some of the caves we have described may not be preserved in the same manner as they are today for future generations.

References

Andreev, S., & Kenderov, L. (2012). Sur une nouvelle espèce du genre Niphargus de la Turquie—Niphargus turcicus n.sp. (Amphipoda, Niphargidae). *Historia Naturalis Bulgarica, 20*, 47–56.

Guéorguiev, B. (2012). Beronaphaenops paphlagonicus, a new anophthalmous genus and species of Trechini (Coleoptera, Carabidae) from Turkey. *ZooKeys, 255*, 73–92.

Tunçkol, B., Aksoy, N., & Yaşayacak, H. (2020). Classification of endemic plants and priority conservation areas in Küre Mountains National Park (Kastamonu section). *Journal of Forestry, 16*(2), 148–160.

Tunçkol, B., Aksoy, N., & Yaşayacak, H. (2022). Küre Dağları Milli Parkı'nın Florası (Kastamonu Bölümü). *Eurasian Journal of Forest Science, 10*(3), 72–107.

Glossary

Active cave	A cave containing a flowing stream.
Alluvium	A general term for clay, silt, sand, gravel, or similar unconsolidated material deposited by a stream in recent geological time.
Aquifer	1. A formation, group of formations, or part of a formation containing sufficient saturated permeable material to yield significant quantities of water to wells and springs. 2. A reservoir of groundwater. 3. Permeable rock that is completely saturated and will yield water to a well or spring.
Architrave	In classical architecture, an architrave, also called an epistyle, is the lintel or beam that rests on the capitals of columns. The term may also be applied to all sides, including the vertical members, of a frame with moldings around a door or window.
Arthropods	Animals with jointed legs and an external skeleton (exoskeleton). The group includes insects, crustaceans, spiders, centipedes, and various other types of animals commonly found in caves.
Bacteria	Free-living organisms often consisting of a single biological cell. They make up a large domain of prokaryotic microorganisms. Typically a few micrometers in length, bacteria were among the first life forms to appear on Earth and are present in most of its habitats.
Bare karst	A type of karst landscape with no soil cover, where dissolution of carbonate rocks to form karst landforms occurs primarily on the exposed bedrock surface basin.
Basin	The area over which groundwater drains to the same point. It may be larger than the associated drainage basin if permeable layers extend beyond the topographic divide.
Bed	1. A layer in sedimentary rock. 2. A sedimentary deposit of relatively small thickness and large area, separated from overlying and underlying deposits by bedding planes.
Bedrock	Solid rock that underlies unconsolidated material.
Biospeleology	The scientific study of subterranean organisms, or cave biology.
Breccia	Angular fragments of rock cemented together by finer-grained materials such as silica, iron minerals, and calcite to form a new rock.
Bronze Age	A historical period lasting from about 3300 BC to 1200 BC. It is characterized by the use of bronze and the use of writing in some regions.
Calcareous	1. Containing calcium carbonate. 2. Describing a rock containing calcium carbonate.
Calcite	The more common and stable mineral form of calcium carbonate. It is the dominant component of all limestones and, due to its dissolution and reprecipitation by natural waters at normal temperature, is the most common form of limestone.

A. Yamaç, E. Tok, *Caves and Karst of Küre Mountains National Park, Turkey*, Cave and Karst Systems of the World,
https://doi.org/10.1007/978-3-031-80142-6

Calcium carbonate — A naturally occurring compound with the chemical formula $CaCO_3$. It occurs commonly as the mineral calcite, and less commonly as aragonite, and is the major constituent of carbonate rocks, including limestone and marble.

Captorhinidae (also known as cotylosaurs) — `An extinct family of tetrapods, usually regarded as primitive reptiles, known from the Late Carboniferous to the Late Permian. They had a cosmopolitan distribution throughout Pangea.

Cave ecology — The study of the interaction between cave organisms and their environment.

Cave system — 1. An underground network of passages, chambers, or other cavities. 2. The caves in a given area that are hydrologically connected, whether continuous or discontinuous from a single opening.

Chalcolithic (Copper Age) — An archaeological period characterized by the increasing use of smelted copper. It followed the Neolithic and preceded the Bronze Age. In the Ancient Near East, the Chalcolithic began in the late fifth millennium BC and lasted for about a millennium before giving way to the Early Bronze Age.

Chemotroph — A class of organisms that derive their energy from the oxidation of inorganic molecules such as iron and magnesium. The most common type of chemotroph is prokaryotic and includes both bacteria and fungi.

Claystone — A compacted, non-siliceous, fine-grained sedimentary rock composed predominantly of clay-sized particles (less than 4 mm in grain size).

Closed depression, closed basin — 1. Any karst depression with internal drainage, including dolines, uvalas, poljes, cockpits, and all types of blind karst valleys, both small and large. 2. A general term for any closed topographic basin without external drainage, regardless of origin or size.

Closed karst — Karst topography that is completely buried by relatively younger post-rock or sediments and is not part of the contemporary landscape.

Collapse sinkhole — 1. A variety of closed depressions formed by the collapse of rock over an existing cave passage or chamber. 2. A closed depression formed by the collapse of the roof of a cave.

Column — An underground dripstone formation formed by the union of a stalactite and a stalagmite in a cave.

Conglomerate — Rock consisting of large, well-rounded, water-worn particles.

Covered karst — A fossil or currently developing karst underlain by surficial deposits or other rocks, which may produce surface landforms that reflect subsurface karstification; as contrasted with bare karst, which is soil-free.

Cretaceous — The third of the three periods of the Mesozoic Era. It began 146 Ma ago and ended 65.5 Ma ago. It is characterized by the mass extinction of many invertebrate and vertebrate species.

Crustaceans — The large class of animals that includes lobsters, crabs, amphipods, isopods, and many similar forms. Crustaceans typically live in water and have many articulated appendages, segmented bodies, and hard exoskeletons.

Curtain — 1. Sinuous, thin sheet of dripstone formed on the roof or walls of a cave. 2. A wavy or folded sheet of flowstone hanging from the roof or projecting from the wall of a cave; often translucent and resonant.

Dendrochronology Scientific method of dating tree rings to the exact year they were formed in a tree. As well as dating them, this can give data for dendroclimatology, the study of climate and atmospheric conditions during different periods in history from the wood of old trees.

Depression A landform sunken or depressed below the surrounding area. Depressions form by various mechanisms.

Doline; sinkhole A basin or funnel-shaped hollow in limestone, ranging in diameter from a few meters up to a kilometer and in depth from a few to several hundred meters. A distinction can be made by direct solution of the limestone surface zone (solution dolines) and those formed by collapse over a cave (collapse dolines).

Dolomite The pure mineral dolomite has the composition $CaMg(CO_3)_2$ and has properties very similar to those of calcite. The rock dolomite consists mainly of the mineral dolomite, with subordinate calcite, and has properties very similar to those of limestone. The natural dissolution of dolomite is generally slower than that of limestone.

Dolomitic limestone A limestone containing a significant proportion of the mineral dolomite but in which calcite is more abundant (e.g., 10–45% dolomite, 90–55% calcite). Many dolomitic limestones originate as calcite limestone that is subsequently affected by magnesium-rich water that replaces part of the calcite with dolomite.

Drainage basin The land area from which surface runoff drains into a stream channel or system of stream channels, or to a lake, a reservoir, or other bodies of water. In a karst setting, subsurface drainage (internal drainage) may have boundaries defined on the basis of comprehensive groundwater tracing studies.

Dripstone Calcium carbonate deposited from water dripping from the ceiling or wall of a cave; commonly refers to the rock in stalactites, stalagmites, and other similar speleothems; in some places composed of aragonite or gypsum.

Dye test Determination of the direction and rate of flow of streams by marking them with dye at the infiltration area and then identifying and timing the reappearance of color at lower-lying springs, in river beds, and elsewhere in a cave system.

Ecosystem A system that environments and their organisms form through their interaction. The biotic and abiotic components are linked together through nutrient cycles and energy flows.

Endemism State of a species only being found in a single defined geographic location. Organisms that are indigenous to a place are not endemic to it if they are also found elsewhere.

Eocene Palaeogene epoch which began at the end of the Palaeocene (55.8 Ma) and ended at the beginning of the Oligocene (33.9 Ma). It is noted for the expansion of mammalian stocks (horses, bats, and whales appeared during this epoch), and the local abundance of nummulites (marine protozoans of the Foraminiferida).

Epikarst; epikarst zone A relatively thick portion of bedrock that extends from the base of the soil zone and is characterized by extreme fracturing and enhanced solution. It is separated from the phreatic zone by an inactive, relatively waterless interval of bedrock that is locally breached by vadose percolation.

Fault Approximately plane surface of fracture in a rock body, caused by brittle failure, and along which observable relative displacement has occurred between adjacent blocks. Most faults may be broadly classified according to the direction of slip of adjacent blocks into dip-slip, strike-slip, and oblique-slip varieties.

Fauna	All of the animal life present in a particular region or time. Zoologists and paleontologists use fauna to refer to a typical collection of animals found in a specific time or place.
Flora	All the plant life present in a particular region or time, generally the naturally occurring (indigenous) native plants.
Flowstone	Deposits of calcium carbonate, gypsum, and other mineral matter which have accumulated on the walls or floors of caves at places where water trickles or flows over the rock. Layered deposits of calcium carbonate precipitated on rocks from water trickling over them.
Flysch	Sedimentary facies term used to describe a thick succession of redeposited, deep-sea, clastic material of synorogenic character.
Foraminiferida (informally foraminifera)	Order of testate, amoeboid protozoa in which the cell is protected by a test, consisting of one to many chambers, whose structure and composition is of great importance in foraminifera classification.
Formation	Series of rock layers that were deposited during the same geological period and exhibit similar characteristics.
Fossil cave	A fossil cave is an underground cavity that is formed when a carbonate succession was undergoing karstification but subsequently buried. Most fossil caves have been infilled by younger sediments.
Frieze	In classical architecture, the frieze is the wide central section of an entablature and may be plain in the Ionic or Doric order, or decorated with bas-reliefs.
Frugivore	An animal that thrives mostly on raw fruits or succulent fruit-like produce of plants such as roots, shoots, nuts, and seeds.
Genus (pl. genera)	A taxonomic rank used in the biological classification of living and fossil organisms as well as viruses. In the hierarchy of biological classification, genus comes above species and below family.
Geomorphology	The science of the origin and evolution of land forms.
Gondwana	Former supercontinent of the southern hemisphere from which South America, Africa, Madagascar, India, Sri Lanka, Australia, New Zealand, and Antarctica are derived. Their earlier connection explains why related groups of plants and animals are found in more than one of the now widely separated southern land masses.
Graben	A depression formed by a fault block moving downward on the two bounding faults.
Grade	1. Inclination or slope. 2. The class of a cave survey on the basis of the precision of the instruments and the accuracy of the methods.
Guano	An accumulated deposit of animal excrement. In caves it is most commonly associated with bat colonies. Consisting mainly of phosphates and nitrates it is valued as a fertilizer or an ingredient of explosives and has commonly been mined.
Helictite	Generally small variety of stalactitic calcite growth that is twisted and contorted with no apparent regard for gravity. Helictites form on cave walls, ceilings, and stalactites. The helictite shape is created by crystal lattice distortion

and crystal form changes within the calcite, but what causes these is uncertain.

Hibernation — A state of minimal activity and metabolic depression undergone by some animal species during winter months. Hibernation is a seasonal heterothermy characterized by low body temperature, slow breathing and heart rate, and low metabolic rate.

Hydrology — The study of the hydrologic (water) cycle. While it involves aspects of geology, oceanography, and meteorology, hydrology emphasizes the study of bodies of surface water on land and how they change with time.

Iconoclasm — Two periods in the history of the Byzantine Empire when the use of religious images or icons was opposed by religious and imperial authorities within the Ecumenical Patriarchate. The First Iconoclasm occurred between about 726 and 787, while the Second Iconoclasm occurred between 814 and 842.

Igneous — Applied to one of the three main groups of rock types (igneous, metamorphic, and sedimentary), to describe those rocks that have crystallized from magma.

Impermeable — A characteristic of some geologic material that limits their ability to transmit significant quantities of water under the pressure differences ordinarily found in the subsurface.

Insectivore — An animal that feeds on insects, worms, and other invertebrates.

Jurassic — One of the three Mesozoic periods. It lasted from 199.6 to 145.5 Ma, following the Triassic and preceding the Cretaceous with clays, calcareous sandstones, and limestones being the most common rock types. Brachiopods, bivalves, and ammonites are abundant fossils, along with many other invertebrate stocks.

Karst — A terrane, generally underlain by limestone or dolomite, gypsum, halite, and other soluble rocks in which the topography is chiefly formed by the dissolving of rock, and which may be characterized by sinkholes, sinking streams, closed depressions, subterranean drainage, and caves. The term karst unites specific morphological and hydrological features in soluble rocks.

Karstification — The processes of solution and infiltration by water, mainly chemical but also mechanical, whereby the surface features and subterranean drainage network of a karstland are developed to form a karst topography, including such surface features as dolines and karrens and such subsurface features as caves and shafts.

Lapiés — Term for a region with outcrops of small regular pillars, cones, or blocks of carbonate rock.

Laurasia — The northern continental mass produced in the early Mesozoic by the initial rifting of Pangaea along the line of the North Atlantic Ocean and Tethys. Laurasia included what was to become North America, Greenland, Europe, and Asia, while the large, southern, continental mass (called Gondwana) was later divided into South America, Africa, India, Australia, and Antarctica.

Liassic — Former name for what is now called the Early Jurassic Epoch, dated at 199.6–175.6 Ma.

Limestone — Sedimentary rock containing at least 50% calcium carbonate by weight. The purer limestones consist almost entirely of calcite; less pure rocks may be referred to as, for example, muddy limestone. Some limestones are porous with diffuse permeability; these rarely become truly cavernous, though some fissure flow may occur. Where groundwater flow in less porous rocks is restricted to bedding-related fissures and secondary fractures, it can, even when moving very slowly, corrode the almost entirely soluble rock and lead to true cave development.

Lithology — The physical characteristics of a rock, including composition, grain size, texture, degree of cementation (or lithification), and structure, which determine the rock type.

Lithosphere — The upper (oceanic and continental) layer of the solid Earth, comprising all crustal rocks and the brittle part of the uppermost mantle. It comprises numerous blocks, known as tectonic plates, which have differential motions giving rise to plate tectonics. Its thickness is variable, from 1 to 2 km at mid-oceanic ridge crests, but generally increasing from 60 km near the ridge to 120–140 km beneath older oceanic crust.

Longitudinal fault — A fault having the same direction of strike as the surrounding strata.

Longitudinal section — A section along the length of a cave passage or chamber or combination of these, or along a survey traverse in a cave.

Marble — Limestone metamorphosed, recrystallized, and hardened by heat and pressure.

Marl — A sediment typically found interbedded with purer ooze in beds up to 1.5 m thick. It is 30% clay and 70% microfossils, at least 15% of its volume being siliceous microfossils.

Mesozoic — The middle of three eras that constitute the Phanerozoic period of time. The Mesozoic was preceded by the Palaeozoic Era and followed by the Cenozoic Era. It began with the Triassic approximately 251 Ma ago and ended around 65.5 Ma at the start of the Cenozoic. The Mesozoic comprises the Triassic, Jurassic, and Cretaceous Periods.

Miocene — First of the two epochs of the Neogene Period, extending from the end of the Oligocene, 23.03 Ma ago, to the beginning of the Pliocene, 5.332 Ma ago. Many mammals with a more modern appearance evolved during this epoch, including deer, pigs, and several elephant stocks.

Obruk — Turkish form for a vertical or steep-sided depression or shaft in karst, often formed by the collapse of the roof of an underground cave or cavern.

Orogeny (orogenesis) — Mountain formation, especially when a belt of the Earth's crust is compressed by lateral forces to form a chain of mountains. There have been many orogenic episodes in the evolution of the crust, each extending over many millions of years.

Paleogene (or Palaeogene) — A geologic period and system that spans 43 million years from the end of the Cretaceous Period 66 million years ago to the beginning of the Neogene Period 23.03 Mya. It is the beginning of the Cenozoic Era of the present Phanerozoic Eon.

Paleokarst — 1. A karstified rock or area that has been buried by later sediments 2. A decoupled contemporary system that has experienced tectonic subsidence and lies unconformably beneath clastic cover rocks 3. A karst formed in the past under an earlier erosion cycle and often in remote geological times. The karst is preserved by burial or suspension of karstification processes. 4. A karstified surface and the karst features associated with it, such as caves, which have been buried by younger rocks.

Paleolithic (or Palaeolithic) — A period in human prehistory that is distinguished by the original development of stone tools, and which represents almost the entire period of human prehistoric technology. It extends from the earliest known use of stone tools by hominins, c. 3.3 million years ago, to the end of the Pleistocene, c. 11,650 BP.

Paleomagnetism	Natural remanent magnetization preserved in rock sequences. During rock deposition magnetic minerals are aligned according to the direction and polarity of the earth's contemporary magnetic field. After movement of the magnetic poles, or periodic reversals of polarity, the remanent magnetization is preserved in the rocks and may be measured to aid identification of stratigraphical units and to assess their relative ages.
Paleontology	The study of life in past geologic time, based on fossil plants and animals and including phylogeny, their relationships to existing plants, animals, and environments, and the chronology of the Earth's history.
Paleozoic (or Palaeozoic)	The first (542–251 Ma) of the three eras of the Phanerozoic. During the Paleozoic, two major orogenies occurred: the Caledonian during the Lower Palaeozoic, and the Variscanin during the Late Palaeozoic times. The faunas of the Palaeozoic are noted for the presence of many invertebrate organisms.
Pangea	A single supercontinent which came into being in late Permian times and persisted for about 40 million years before it began to break up at the end of the Triassic Period.
Passage	Broadly, a passage is any negotiable part of a cave system, though the usage is commonly restricted to those elements that tend toward the horizontal rather than vertical or subvertical sections. Cave passages vary in size and shape, with the latter relating to the mode of origin and providing evidence of the nature of cave development mechanisms.
Percolation	Groundwater moving slowly through the micro-fissure network of a limestone, most of which eventually joins a major cave conduit and flows more rapidly. In most environments, percolation water enters the limestone through a soil cover. It is therefore high in carbon dioxide and has a major influence on limestone dissolution and later redeposition of calcite speleothems.
Permeability	It is a measure of the ability of a porous rock to allow fluids to pass through it. The permeability of a medium is related to the porosity, but also to the shapes of the pores in the medium and their level of connectedness
Petrography	The science of describing and identifying rocks.
Phreatic zone	The zone of saturated rock below the water table, within which all conduits and sub-conduits are water filled.
Phylogenetics	Relating to the evolutionary development and diversification of a species or group of organisms, or of a particular feature of an organism.
Phylum	In biology, a phylum is a level of classification or taxonomic rank below kingdom and above class.
Plate tectonics	The unifying concept that has drawn continental drift, sea-floor spreading, seismic activity, crustal structures, and volcanic activity into a coherent model of how the outer part of the Earth evolves. The theory proposes a model of the Earth's upper layers in which the colder, brittle, surface rocks form a shell (the lithosphere) overlying a much less rigid asthenosphere.The shell comprises several discrete, rigid units (tectonic plates) each of which has a separate motion relative to the other plates. Many features of the Earth's history are explicable within this model which has served as a unifying hypothesis for most of the Earth sciences.
Pleistocene	An epoch of the Quaternary period, after the Pliocene of the Tertiary and before the Holocene; also the corresponding worldwide series of rocks. It began two million years ago and lasted until the

start of the Holocene some 8000 years ago.

Polje — A large, flat floored depression in karst limestone, whose long axis is developed parallel to major structural trends and can reach tens of kilometers in length. Drainage may be by either surface watercourses or swallow holes. Commonly the ponors cannot transmit flood flows, so many poljes turn into wet-season lakes.

Pothole — A single shaft, or an entire cave system that is dominantly vertical. It is also used to describe a single erosional bowl or moulin, rounded mainly by the swirling current, in a stream bed.

Pre-Pottery Neolithic — An archaeological period that represents the early Neolithic in the Levantine and upper Mesopotamian region of the Fertile Crescent, dating to c. 12,000—c. 8500 years ago (10000–6500 BCE). It succeeds the Epipalaeolithic (also called Mesolithic) and succeeded by the Pottery Neolithic.

Pseudokarst — A landscape containing karst-like features such as caves and dolines, but not formed by bedrock dissolution as in true karst. Pseudokarst embraces volcanic landscapes with lava caves, cryokarst, or thermokarst formed by ground-ice melting in a permafrost environment.

Radioactive dating (Radiometric dating) — A technique which is used to date materials such as rocks or carbon. The method compares the abundance of a naturally occurring radioactive isotope within the material to the abundance of its decay products, which form at a known constant rate of decay. The technique is now the principal source of information about the absolute age of rocks and other geological features, including the age of fossilized life forms or the age of Earth itself, and can also be used to date a wide range of natural and man-made materials. Among the best-known techniques are radiocarbon dating, potassium–argon dating, and uranium–lead dating.

Rift — It is a linear zone where the lithosphere is being pulled apart and is an example of extensional tectonics. Typical rift features are a central linear downfaulted depression, called a graben, or more commonly a half-graben with normal faulting and rift-flank uplifts mainly on one side. Where rifts remain above sea level they form a rift valley, which may be filled by water forming a rift lake.

Sandstone — Sedimentary rock type, formed of a lithified sand, comprising grains between 63 micron and 1000 micron in size, bound together with a mud matrix and a mineral cement formed during burial diagenesis.

Sediment — Material recently deposited by water, ice, or wind, or precipitated from water.

Shaft — 1. Vertical, or steeply inclined, sections of a cave passage, of enormously varied size. 2. A cylindrical tube generally steep sided, which forms by solution and (or) collapse. 3. A vertical passage in a cave. 4. A vertical and usually large diameter hole penetrating geologic formations for access of subsurface points.

Shale — Fine-grained, fissile, sedimentary rock composed of clay-sized and silt-sized particles of unspecified mineral composition.

Single rope technique — The practice of climbing up and down ropes with the help of ascenders and descenders. Abbreviation: SRT.

Sinkhole — A point where a stream or river disappears underground. The sinking water may filter through a choke that excludes cavers, or may flow into an open horizontal cave or vertical shaft, and while active all of these may be termed sinkholes. The flow of water may be very small, but in full flood many sinkholes swallow flows of tens of cubic meters per second.

Siphon — Synonym for a sump, gallery in the form of an inverted "U" with water moving only under pressure when the siphon has completely filled up; the water head at the input end being higher than at the drainage point. In speleology, a cave passage in which the ceiling dips below a water surface.

Species — Largest group of organisms in which any two individuals of the appropriate sexes or mating types can produce fertile offspring, typically by sexual reproduction. It is the basic unit of classification and a taxonomic rank of an organism, as well as a unit of biodiversity.

Speleogenesis — Although the term literally means the birth, origin, or mode of formation of caves, the full extent of speleogenesis includes all the changes that take place between the inception and the eventual destruction of an underground drainage system. It includes development phases during which the active drainage voids are too small to be considered caves as normally defined, as well as phases when the cave no longer functions as a drain, is enlarging only by collapse, and, eventually, is being totally removed.

Speleology — The scientific study, exploration, and description of caves, including aspects of sciences, such as geomorphology, geology, hydrology, chemistry, and biology, and also the many techniques of cave exploration.

Speleothem — General term for all cave mineral deposits, embracing all stalactites, stalagmites, moonmilk, helictites, flowstones, flowers, etc., and other secondary mineral deposits in caves. Most are formed of calcite whose precipitation processes, related mainly to carbon dioxide levels in the water, are the direct reverse of the dissolution of limestone.

Stalactite — Speleothem, normally of calcite, formed by dripping water and hanging from a cave roof. Stalactites embrace an enormous variety of sizes and shapes. They form where percolation water seeps from a cave ceiling and becomes saturated with respect to calcite due to loss of carbon dioxide into the cave air.

Stalagmite — Speleothem, normally of calcite, formed by upward growth from a cave floor, and therefore the complement of a stalactite. Stalagmites form when dripwater that is still saturated falls from a cave roof or stalactite and, when or after it lands, loses more carbon dioxide to the cave air, causing precipitation of calcite.

Strike-slip fault — A fault in which the major displacement is horizontal and parallel to the strike of a vertical or subvertical fault plane. Localized zones of deformation due to pressures and tensions across the fault occur at bends in the fault strike and give rise to the formation of pull-apart basins.

Subduction zone — The zone, at an angle to the surface of the Earth, down which a

	lithosphericplate descends. Most present-day subduction zones extend from trenches on the ocean floor, from where a zone of earthquake hypocenters extends.
Survey	In caving, the measurement of directions and distances between survey points and of cave details from them, and the plotting of cave plans and sections from these measurements either graphically or after computation of coordinates.
Taxa	In biology, a taxon is a group of one or more populations of an organism or organisms seen by taxonomists to form a unit. Although neither is required, a taxon is usually known by a particular name and given a particular ranking, especially if and when it is accepted or becomes established.
Tectonism	Deformation within the Earth's crust, and its consequent structural effects.
Tethys Sea (Neotethys)	The sea that more or less separated the two great Mesozoic supercontinents of Laurasia (in the north) and Gondwana (in the south). That land bridges between the two supercontinents existed for much of the Mesozoic is attested to by the cosmopolitan character of dinosaur faunas.
Thrust; thrust fault	A generally gently dipping or subhorizontal reverse fault plane where the relative movement has been essentially horizontal, with one rock sequence being pushed across and above another.
Triassic	The earliest (251–199.6 Ma) of the three periods of the Mesozoic Era. As a result of the mass extinctions of the late Palaeozoic, Triassic communities contained many new faunal and floral elements. Among these were the ammonites, modern corals, various mollusks, the dinosaurs, and certain gymnosperms.
Troglobite	A creature that is fully adapted to life in total darkness and can only complete its life cycle underground. Many troglobitic species are adapted in some way to living in a totally dark environment.
Troglomorphy	The physical characteristics of a troglobite or stygobite, e.g., reduced eyes and pigment, elongated appendages, and well-developed tactile and olfactory organs.
Troglophile	"Cave lover." An animal that can complete its life cycle in caves, but may also do so in suitable habitats outside caves.
Trogloxene	"Cave visitor." An animal that habitually enters caves, but must return periodically to the surface for certain of its living requirements, usually food.
Tumulus (pl.: tumuli)	A mound of earth and stones raised over a grave or graves. Tumuli are also known as barrows, burial mounds, and may be found throughout much of the world.
Twilight zone	The area of a cave where light penetrating through the entrance is sufficient to permit human vision.
Vadose zone	The zone between the land surface and the water table.
Vertebrate	An animal with a backbone. The group includes fishes, amphibians, reptiles, birds, and mammals. Some amphibians and fishes live permanently in caves.
Vertical cave	A vertical passage within a cave system, formed along joints by which underground water courses are transferred from a higher to a lower bedding plane. They may become transformed into vertical shafts by sufficiently uniform dissolution as to create a rounded vertical passage.
Water table	The top surface of a body of slowly moving groundwater that fills the pore spaces within a rock mass. Above it lies the freely draining vadose zone, and below it lies the permanently saturated phreas. Individual cave conduits may be above or below the water table, and therefore either vadose or phreatic, and the water table cannot normally be related to them.

BCRA Cave Map Survey Grades

You will see a note like "BCRA 3C" on almost all the cave maps found in this book. This is a cave survey rating system which was created in the 1960s by the British Caving Research Association (BCRA) to provide a standard for cave map drawing. Over the years, BCRA grades were spread over the globe and often used internationally. Depending on the developing technology and the problems observed in the old system, BCRA has revised the grades several times over the past few years.

For interested readers, the following tables present the grades, the accuracy of details, additional information, and an explanation which helps in understanding the meaning of the tables (BCRA, n.d.; Häuselmann, 2011).

Table 1 BCRA gradings for a cave line survey

Grade 1 Sketch of low accuracy where no measurements have been made
Grade 2 (use only if necessary, see note 7) May be used, if necessary, to describe a sketch that is intermediate in accuracy between Grades 1 and 3
Grade 3 A rough magnetic survey. Horizontal and vertical angles measured to ±2.5°; distances measured to ±50 cm; station position error less than 50 cm.
Grade 4 (use only if necessary, see note 7) May be used, if necessary, to describe a survey that fails to attain all the requirements of Grade 5 but is more accurate than a Grade 3 survey.
Grade 5 A Magnetic survey. Horizontal and vertical angles measured to ±1°; distances should be observed and recorded to the nearest centimeter and station positions identified to less than 10 cm.
Grade 6 A magnetic survey that is more accurate than Grade 5 (see note 5).
Grade X A survey that is based primarily on the use of a theodolite or total station instead of a compass (see notes 6 and 10 below).

Notes

1. The above table is a summary and is intended only as an *aide memoire*; the definitions of the survey grades given above must be read in conjunction with these notes.
2. In all cases, it is necessary to follow the spirit of the definition and not just the letter.
3. To attain Grade 3, it is necessary to use a clinometer in passages having appreciable slope.
4. To attain Grade 5, it is essential for instruments to be properly calibrated, and all measurements must be taken from a point within a 10 cm diameter sphere centered on the survey station.
5. A Grade 6 survey requires the compass to be used at the limit of possible accuracy, i.e., accurate to ±0.5°; clinometer readings must be to the same accuracy. Station position error must be less than ±2.5 cm, which will require the use of tripods at all stations or other fixed station markers ("roofhooks").
6. A Grade X survey must include on the drawing notes descriptions of the instruments and techniques used, together with an estimate of the probable accuracy of the survey compared with Grade 3, 5, or 6 surveys.
7. Grades 2 and 4 are for use only when, at some stage of the survey, physical conditions have prevented the survey from attaining all the requirements for the next higher grade and it is not practical to re-survey.
8. Caving organizations are encouraged to reproduce Tables 1 and 2 in their own publications; permission is not required from BCRA to do so, but the tables must not be reprinted without these notes.
9. Grade X is only potentially more accurate than Grade 6. It should never be forgotten that the theodolite/Total Station is a complex precision instrument that requires considerable training and regular practice if serious errors are not to be made through its use!
10. In drawing up, the survey coordinates must be calculated and not hand-drawn with scale rule and protractor to obtain Grade 5.

Table 2 BCRA gradings for recording cave passage detail

Class A All passage details based on memory.
Class B Passage details estimated and recorded in the cave.
Class C Measurements of detail made at survey stations only.
Class D Measurements of detail made at survey stations and wherever else needed to show significant changes in passage dimensions.

Notes

1. The accuracy of the detail should be similar to the accuracy of the line.
2. Normally only one of the following combinations of survey grades should be used: 1A, 3B or 3C, 5C or 5D, 6D, XA, XB, XC, or XD.

A. Yamaç, E. Tok, *Caves and Karst of Küre Mountains National Park, Turkey*, Cave and Karst Systems of the World,
https://doi.org/10.1007/978-3-031-80142-6

References

BCRA. (n.d.). Cave Surveying Group (retrieved Apr. 8th. 2023). Retrieved from https://www.bcra.org.uk/surveying/

Häuselmann, P. (2011). UIS mapping grades (technical note). *International Journal of Speleology, 40*(2), IV–VI.

Index

A. Yamaç, E. Tok, *Caves and Karst of Küre Mountains National Park, Turkey*, Cave and Karst Systems of the World,
https://doi.org/10.1007/978-3-031-80142-6

The manufacturer's authorised representative in the EU is Springer Nature Customer Service Centre GmbH, Europaplatz 3, 69115 Heidelberg, Germany. If you have any concerns regarding our products, please contact ProductSafety@springernature.com

Printed and bound by CPI Group (UK) Ltd, Croydon, CR0 4YY

30/09/2026

02225516-0014